设计的共性——从产品、建筑到城市规划

DESIGNING OBJECTS BUILDINGS CITIES

[意] 斯德凡罗·德·安格理斯 (Stefano de Angelis)
著
[意] 玛利亚·玛扎 (Maria Mazza)　　温为才

陈华　译

内容简介

平面设计、产品设计、建筑设计、城市设计，有其边界，亦有其融合及共性。设计要解决当今社会发展的问题，是一项系统的设计任务，今天的设计师要有跨界的能力，因此理解各类设计的共性至关重要。本书的著者来自瑞士、意大利、中国，三位著者都有高校执教的背景，亦有丰富的实践经验。作者从不同的文化视角切入设计的基本问题，探讨设计的共性要素，是一个非常有趣且有启发性的过程。本书用最简洁的语言及图表，简明扼要地阐述了设计的要素，设计的方法及案例，在未来建筑及城市设计的零能量消耗这一个大众关注的热点上提出了大量创造性的系统方法，希望可以和广大读者分享世界前沿的绿色设计理论与方法。本书不仅是一本广大设计学子及设计师可以仔细品读的佳作，也可为广大从事城市建设及城市管理的人员提供大量的有益建议。

图书在版编目（CIP）数据

设计的共性：从产品、建筑到城市规划：汉英对照/（意）斯德凡罗·德·安格理斯，（意）玛利亚·玛札，温为才著；陈华译. -- 北京：电子工业出版社，2017.8
ISBN 978-7-121-31897-9

Ⅰ. ①设… Ⅱ. ①斯… ②玛… ③温… ④陈… Ⅲ. ①城市规划－建筑设计－汉、英 Ⅳ. ①TU984

中国版本图书馆CIP数据核字(2017)第133728号

策划编辑：赵玉山
责任编辑：张小乐
印　　刷：北京盛通印刷股份有限公司
装　　订：北京盛通印刷股份有限公司
出版发行：电子工业出版社
北京市海淀区万寿路173信箱　邮编100036
开　　本：880×1230　1/16　印张：16　字数：410千字
版　　次：2017年8月第1版
印　　次：2017年8月第1次印刷
定　　价：99.00元

凡所购买电子工业出版社图书有缺损问题，请向购买书店调换。若书店售缺，请与本社发行部联系，联系及邮购电话：(010)88254888，88258888。
质量投诉请发邮件至zlts@phei.com.cn，盗版侵权举报请发邮件至dbqq@phei.com.cn。
本书咨询联系方式：(010) 88254556，zhaoys@phei.com.cn。

Preface to the Chinese edition

The Chinese edition of "Designing objects buildings cities" is a history of cultures and friendship.

Design has always been our passion.

The space around us, buildings, streets and cities, are our laboratory: every square, every façade and every building is the pretext for a series of critical remarks aimed at searching for the hidden potential, what could be changed to improve the quality of the space in which we live.
We believe that the reason why a large part of the possible improvement of quality of life is "thrown to the winds" in our cities and in our lives is due to the inability to focus on basic questions, on the whys.

The basic concepts of design are, in fact, systematically "watered down" during the teaching and the transmission of information in general, with the result that the essence of things becomes difficult to be found for the majority of people, including designers.

A few years ago, we wrote two small books on the design of buildings and cities, published as e-books, with simple and clear ground rules, and intended both for students and anyone interested in understanding what are buildings and cities today.

Four years ago, we had the chance to meet in Italy with Stone Wen, designer and professor at Wuyi University, during the inauguration of an exhibition of sculptures, at which Stone and Maria had been invited to submit their works: it was the start of a great friendship.
We accepted with enthusiasm when Stone Wen invited us to present our zero-consumption glass building at his university in Guangdong.

The first two conferences in Guangdong were followed by conferences on the future of cities and buildings in six other Chinese universities, as well as several workshops held jointly with Stone Wen, which cemented our friendship: it was a wonderful experience in this huge and dynamic country that is open to the future.

Hence, the request from Stone Wen to further refine our basic texts and to publish in China a book that summarizes the content of our conferences and our workshops, i.e. our planning philosophy.

Our "recipes", impregnated by the comparison between our two reference cultures, Germanic and Latin, have been enriched by the millennial Asian wisdom through our more recent contact with Chinese culture.
With Stone Wen, we were received in Beijing by director Mr. Tan Haiping, to whom we are very grateful for confirming to us, after a brief presentation of the project, that the book was going to be published by PHEI.

The Chinese edition of "Designing objects buildings cities" was born:
thank you Stone!

Stefano and Maria

中文版前言

中文版《设计的共性 —— 从产品、建筑到城市规划》是文化碰撞及友谊的硕果。

设计是我们的激情所在。

我们置身于空间中，从建筑到街道、城市，都是我们的实验田。每个广场，每个立面，每幢建筑都将面临一系列苛刻的言论，这些言论志在寻求完善的可能性，将提高我们的居住空间品质。我们相信，由于我们缺乏对一些基本问题，以及导致这样问题源由的把握能力，因此大多数改善我们生活品质的可能提案，被扔置了身后。

设计的基本概念，从本质上说，在教育及传递过程中被系统地削弱了，其结果是对于大多数人，包括设计师，这些事物的本质变得难以寻觅。

好几年前，我们写了两本有关设计建筑及城市的小册子，并出版了电子书，以简洁而清晰的法则告诉那些对今天建筑及城市充满兴趣的学生及大众。

四年前，在意大利我们有机会认识了温为才设计师（五邑大学副教授），在一次雕塑展览的开幕式上，温博士及Maria被邀请展出他们的作品，这是我们珍贵友谊的开始。期间我们接受了温博士的邀请，前往广东展示了我的零能量消耗的玻璃建筑。

在广东我们举办了两个讲座，主题是未来的建筑与城市，在中国其他的六个城市，我们与温博士创办了几个工作坊，这无疑加深了我们的友谊。来到在一个如此巨大且充满活力、奔向未来的国家，我们的体验是非凡的。

因此温博士提议，何不基于我们的讲座内容及工作坊、设计哲学，进一步完善我们的内容，在中国出版一本书呢？

我们的书，孕育在德国及拉丁民族的文化背景之上，通过我们最近对中国文化的接触，现在又加入来自亚洲千年文化的智慧。通过温博士，我们在北京与谭海平社长见面。在经过一番简要的陈述后，谭社长确认，本书可以在电子工业出版社出版，在此，我们表示深深的谢意。

中文版《设计的共性——从产品、建筑到城市规划》得以顺利发行，
感谢温博士！

Stefano and Maria 夫妇

简介

设计是什么？
设计是改造世界的工具。

我们思考如何开拓周围的空间，反过来，生存的空间也制约着我们的思考。设计就是在生活中不断开拓创新的过程。

近25年来，我们设计了很多建筑，甚至城市的整个片区。我们一直践行着我们的梦想：为人类创造更好的生活空间，让人们在城市中幸福地生活，让未来更加美好。

设计师们经常从图像、概念和符号中汲取灵感，再以同样的方式来有效地表达他们的设计理念。

本书将设计和设计的流程归纳成一些简单的概念和精要的知识点，阅读本书就像一场惬意的旅行。

旅行的第一站我们先了解一些设计的基本原理，在无限微观到无限宏观，这些原理都不会改变，接着是建筑设计的一系列思考，最后讨论的是城市变迁的必然性。

从微观到宏观，从小至原子到大至大都市，我们向读者呈现了这些不变的法则。同时你也会发现只有注重协调性，具备职业道德，充满敬畏的设计才能更好地打造我们现在和将来的生活。

Stefano de Angelis 及 Maria Mazza

Introduction

What does it mean to design?
Design is the tool to improve the world.

With our minds, we create the space around us and likewise the space in which we live shapes our minds.
Designing is to create, is to live.

In these 25 years of professional activity, we have designed buildings and parts of cities chasing our dream of creating better spaces and smarter locations for the life of a new man, for a better future: to live happily in buildings and cities is definitely possible and also necessary.

Designers feed on images, concepts, symbols and use them to communicate ideas effectively.

We propose a journey through design and its processes, summarized in simple concepts and targeted information.

This journey of three stages starts with the concepts that underlie the design which, as with the universe, do not change from the infinitely small to the infinitely large and then continues through a series of reflections on building design and the necessary transformation of our cities.

We would like to accompany the readers, those who approach these issues along the path from the infinitely small to the infinitely large, from the atom to the megalopolis, on a journey to discover that the rules do not change and that a design based on coherence, ethics and respect can lead to improving the lives of today and tomorrow.

Stefano de Angelis and Maria Mazza
The three stages of the journey that proposes Designing

本书的三大部分产品、建筑、城市规划是 DeltaZERO 的设计理念，以及创建者们思考的具体表述。

设计师和建筑师都担负着重要的社会责任：设计师应该利用好资源，以人为本，尊重环境。

书中阐述的原理都非常直接明了，因为现代社会的读者需要快速接收信息。
当今的社会极其复杂多面，决策者们和想要理解本书的读者们必须在短时间内吸收消化大量的信息。

基于以上考虑，书中的内容分三步讲述：从产品设计，到建筑物设计，再到未来城市设计。这些内容综合起来就像复合药片一样可以一口吞下，但却可以让我们长时间回味、思考。

设计项目及设计法则

任何领域，任何规模（小到一个餐叉，大到一个城市）的设计品质都遵从同样的标准要求，都必须考虑时代特征，以人为本，尊重环境。

本书总结了设计的机制和设计的附带影响。

“设计项目及设计法则”这篇内容是作者对设计项目的一般思考总结。

设计无处不在，但好的设计工作并不能轻易完成。设计必须要有职业道德的支撑，普世法则的约束才能为人类做出贡献。

我们需要用一种快速而精准的思维来理解如何做好设计。理解获得一致性设计的流程。坚持使用这些基本法则是获得好的设计品质的最好方式，也是分析设计项目的基本要求，它既非陈词滥调，也非时尚潮流。

objects buildings cities are the application of the **DeltaZERO** design philosophy and some of the reflections of its founders.

Designers, as builders, have important social responsibilities: the designer has to use resources in a coherent manner, respecting people and the environment.

The principles set out in the texts are very direct and concise: the recipient of information today has little time.
The world is extremely complex and multifaceted: those who want to understand and decision makers must be able to digest huge amounts of information in a short time.

From these considerations comes the proposal for a three-step path through design: designing objects, buildings for a livable world, and cities for the next future are synthesized into pills to be swallowed quickly, but nevertheless allowing for long term reflections.

The project and its rules

The design of quality in all areas and at any scale (from the fork to the city) is based on the same rules and on the application of coherence with respect to time, place and man.

The text summarizes the mechanisms of design and its side effects.

The project and its rules are a summary of the thought of Stefano de Angelis and Maria Mazza on the project in general.
The design, a universal act, the principal activity of those who do good, cannot be improvised and must be supported by ethics and universal regulation principles to lead to positive effects for humans.
A pattern of quick and concentrated thinking is proposed to understand how to design well, identifying the processes that form the basis of a “coherent design". Consistent application of the basic rules is the best way to design quality and for analysis of a project beyond the cliches and fashions.

i时代建筑设计

好建筑产出的能量跟它消耗的能量一样多，而且不产生垃圾物，能够保证空气舒适清新，这样的建筑设计集高度的连贯性概念、美学、技术于一体，这才是一项负责任的设计成果。

本书是建筑专业同学的实用手册，对现代建筑设计感兴趣的读者也可将它作为剖析设计流程的好工具。

“**i时代建筑设计**”源于Stefano de Angelis在米兰理工大学开讲两门建筑技术课程。一幢先进、稳固、方便维护、低能耗的建筑设计必须有这些基本理论的支撑。SdA开展系列讲座的目的就是将这些最基本的理论法则打包成极易消化的知识压缩包。

2050城市规划设计（再规划）

城市就是我们这个时代的写照：混乱不堪，污染严重，毫无生机。
但是，到底什么是城市？在不远的将来城市将变成什么样子？它们要如何转变才能迎接未来的挑战？

本书大致罗列了一些分析城区的基本标准，并为改造城市的设计提出了一套可行的指导措施，以期将来人们都拥有更体面的生活，不仅是个体本身，他周围的人们都能拥有更多的成长机会，就像在民主平等的社会，大家都拥有和谐平静的生活。

“**2050城市设计（再设计）**”是Stefano de Angelis和Maria Mazza 关于未来城市一系列讲座的理论成果。

Design **a building** for the iGeneration

A good building nowadays produces as much energy as it consumes, produces no waste, and ensures a comfortable atmosphere: it is the result of a complex and responsible design process, integrating conceptual, aesthetic and technical considerations with a high degree of coherence.
The text is a manual for architecture students but also for anyone interested in understanding the process of designing a contemporary building or for those interested in analyzing a building with a clear and supple verification tool.

Design **a building** for the iGeneration is the translation onto paper of two series of courses in architectural technology held by Stefano de Angelis (SdA) at the Politecnico di Milano. The lecture series was developed by SdA to transmit those basic principles necessary for the design of an advanced, long-lasting, maintenance-friendly and low power consumption building in easily digestible but high conceptual-density pills.

(Re) Designing **towns** for 2050

Cities are the mirror of our times: chaotic, polluted and unattractive.
But what is the city? How will cities evolve in the near future? How will they have to change in order to overcome the difficult challenges of the coming decades?
The text outlines the basic criteria for the analysis of urban land and proposes a set of guidelines for applying measures to design for the necessary transformation of cities, to allow future generations a dignified life rich with opportunities for personal and collective growth: a peaceful life in harmony with the principles of democracy and equality.

(Re) Designing **towns** for 2050 is the transfer onto paper of a series of lectures by Stefano de Angelis and Maria Mazza on the future of cities.

城市将很快成为人们评估环境的参照物。因此，我们应该关注城市，让我们的生活和梦想融入我们的城市，让城市成为我们的欣喜之地。

本书的内容并不是随意的拼凑，也不是生硬的教条，而是一种认识，我们这个飞速变化的世界就像我们的思维一样，是可延伸的，灵活的，势不可挡的。

要找到解决难题的答案，我们就应该摒弃原有的、套用现成配方的这种想法。我们应该学会思考，持着怀疑的态度，不断探索研究。

希望接下来的内容和图片可以抛砖引玉，引起大家的思考，有机会来验证我们坚定的信念，为改善我们的生活贡献一份力量，为那些决定开启艰难旅程来“一探究竟”的人们提供一点支持。

Cities are soon going to be the environment of reference for the entire population of our planet. For this reason, we should pay attention to our cities and make them stimulating places in tune with our lives and our dreams.

Designing **objects buildings cities** does not want to give recipes; it is not based on dogma but on the understanding that our rapidly evolving world is ductile, flexible and unstoppable like the thought.

To direct the thinking towards the solution of complex problems, we have to break away from the belief that we need recipes: thought must be trained toward reflection, doubt, research.

The text and images that follow are intended to be a starting point for reflection, an opportunity for verification of convictions and a contribution to the relentless pursuit of all of us committed to improving our life, of those who have decided to embark on the difficult journey into the “why”.

1 事物的原理

1 TOE (Theory Of Everything)

01 - cavolo romanesco, macro scale

02 - cavolo romanesco, micro scale

1 事物的原理

如果仔细回想我们的日常生活，我们就会发现，不论是我们的生活空间，还是我们生活需要的一切物品，都是为了满足某种特定的需求而设计的，是我们思考和行动的一种结果，也是一项设计项目的成果。

一个好的设计项目，不管它是一副餐叉，一幢大楼，还是一座城市，评判优劣的基本标准却是一样的。设计项目里的元素（形式，范围，规模）虽大相径庭，但它们却遵循同一套标准，演绎着同一个正确的理念。

01 西兰花，宏观比例
02 西兰花，微观比例

西兰花
通过电脑处理的几何图形可以按不同比例不断重复，就像几何中的分形结构，在大自然中，我们也可以找到有同样特征的元素。
无论在宏观还是微观比例下，西兰花的尖点也有同样的形状。

1 TOE (Theory Of Everything)

If we analyze our everyday life, we realize that what surrounds us, everything in which we live and what we live with, was built, designed and responds to specific needs or is the consequence of an action or a thought; it is the result of a project.

The basic rules for a good project are surprisingly the same whether it's a fork, a building or a city.
Elements apparently so different in form and scope, very different in scale, share the rules that lead to a correct conception.

CAVOLO ROMANESCO
Like fractals, geometric representations processed through a computer that have shapes repeated at different scales, we find elements with such characteristics even in nature.
Broccoli crowns have the same shape at both the macro and micro scale.

03 - multiverse, universe, galaxy

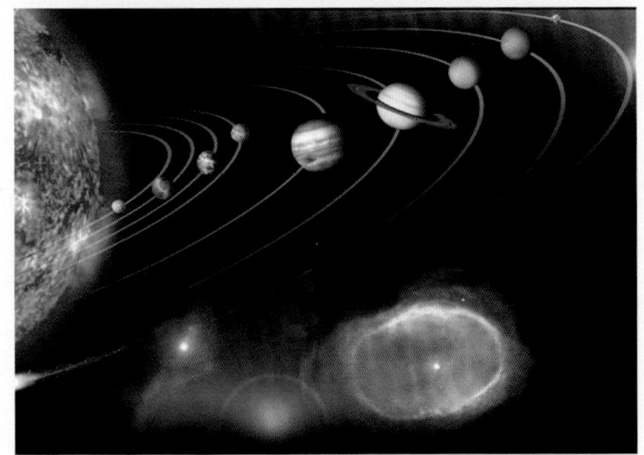
04 - solar system

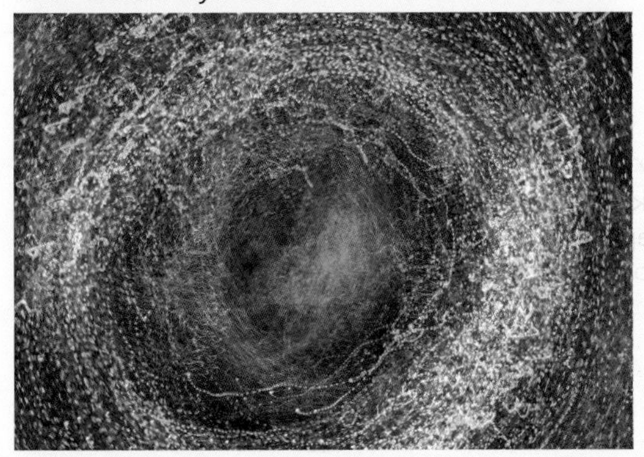
05 - atom

“万物理论”的观点是：小到原子大到超引力，所有的系统和机制都是相似的。同样，设计领域里也是如此：不论何种规模的设计，指导原则却是一样的。不管设计的规格尺寸如何地不同，看待整个项目的思维是一样的。

所有满足人类需求的项目，只要是从职业道德和社会观点来看是正确的，它的内在机制和法则就都是一样的，无论从无穷大还是无穷小的角度来看，这点也不会改变。

03 宇宙群，宇宙，银河系
04 太阳系
05 原子

As in the "Theory of everything" (from the atom to infinity, the forces, systems and mechanisms are similar), so also in design the guiding principle at different scales is always the same. No matter the scale, no matter the size of what you are designing, it is the clarity of thought that determines the project. From the infinitely large to the infinitely small, the mechanisms and rules leading to a project that meets the needs of man and is correct from the point of view of ethics and society are the same.

宇宙群
多个宇宙组成的多重宇宙
多个银河系组成的宇宙
多个太阳系组成的银河系
无数的原子组成太阳系

multiverse - set of universes
universe - set of galaxies
galaxy - set of solar systems
solar system - set of planets
atom - set of atomic particles

2 合理性和独创性
2 rationality and creativity

2 合理性和独创性

一位好的设计师或者是建筑师，不仅应该懂得如何去设计,而且还必须能够全面认识他所设计的东西。
亲身体验建造的过程，有所收获的过程，才能总结出有用的知识来提升整个计划本身。

现如今，一位好的设计师应具备以下几点能力：

- 熟知并能运用当前的先进技术
- 能管理好一个团队
- 对整个项目的经济金融运作有良好的把控能力。

有需求才会有设计，设计的成品有可能是产品，也有可能是建筑物或者是一个城市。成品经过一系列的创新和验证后开始生产，并投放市场销售。

设计过程分为以下几个主要阶段：

- 数据收集
- 分析数据
- 对数据进行整合、过滤
- 验证和修正

设计是一项具有创造力的活动，通过这项活动,一个人或是几个人的创意就能将需求成功转化为物品。

设计过程最大的特点就是要经过多个环节不断地补充，再通过框架环境条件过滤，待时机成熟时才形成产品的确切定义。

2 RATIONALITY AND CREATIVITY

A good designer or architect must not only know how to design, but must also be able to realize what he designs. The experience of building, of achieving, is fundamental to acquiring useful knowledge and improving the quality of the planning process itself.

Nowadays, the good designer :

- knows and can use the technologies currently available
- has the skill to manage a team of people
- knows and respects the economic and financial framework of the project.

The design of a product (be it an object, a building or a city) arises from an assignment that was born from a need, develops through a series of creative and verification phases that end with the start of production and the subsequent commercialization.

The main stages of the design process are :

- data collection
- analysis
- combination and distillation
- verification and correction

Design is a creative act by which one or more minds capable of conceiving ideas succeed in converting requirements into objects.

The process is distinguished by a series of inputs that with a sequence of phases, and passage through the filters of the framework conditions and time, reaches maturity and subsequently the product definition.

设计的过程就是人自己思维里两股势力不断相互扭打的过程：

- 合理性（规则）：原理
- 独创性（构想出新的东西）：创意

对于一项优秀的设计来说，合理性和个人情绪思维应该能做到和谐共处，相互平衡。
原理和规则是设计之身，而灵感创意是设计之魂。没有灵魂的身躯终归也是没有生命的。

06 合理性和独创性

The design process is characterized by the continuous interaction of two forces of man's own mind:

- rationality (rules): principles
- creativity (the ability to conceive the new): idea

For a winning design, both the rational and the emotional thinking need to be in harmony and in balance with each other.
The principles and rules are the body of a good design; the idea is the soul, without which the body does not have any life.

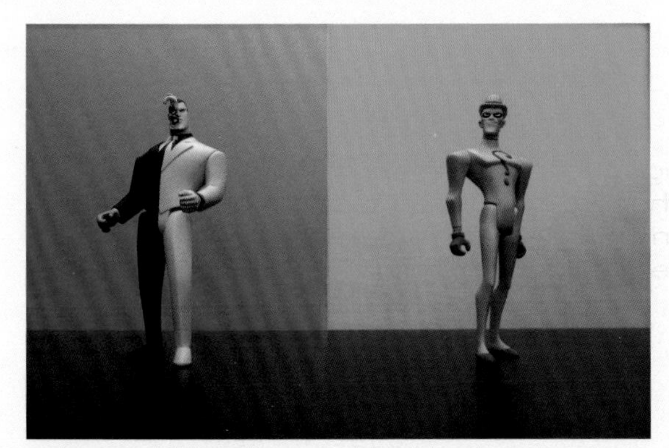

06 - rationality and creativity

3 一致性
3 coherence

3 一致性

内在连贯性是指系统或任何实体内各个部分之间相互独立却又相互联系这样一种关系。
一致性是思想和行为之间逻辑关系的一种表现。
连贯性是任何设计的重要原理之一。
设计项目的理念贯穿设计过程的每个环节，而且很容易从设计成品上识别出来。

07 捷豹C-X17概念车-外观
08 捷豹C-X17概念车-外观

3 COHERENCE

The term coherence refers in general to the connection and interdependence of the parts of a system, an object or any entity.
Coherence indicates a logical relationship between thought and deed, between the design idea and the designed object.
The principle of coherence is one of the cornerstones of design at all scales.
The idea upon which a project is based must guide every choice during the design process and has to stay recognizable in the final product.

07 - Jaguar C-X17 concept - exterior

08 - Jaguar C-X17 concept - exterior

09 - Rolls Royce Phanton - interior

10 - Jaguar C-X17 concept - interior

捷豹C-X17

捷豹C-X17概念车是时尚与高端的标志：高端，积极，富有未来感。如果捷豹C-X17搭配的是09图的内饰（劳斯来斯幻影的内饰），任何人都能马上发现有点不对劲：让人明显感觉到不一致，就是因为没有遵循设计的一致性原则。如果将概念车的现代化高端技术设计理念搭配很古老传统的外观材质就会违背一致性原则。

09 劳斯来斯-幻影-内饰
10 捷豹C-X17 概念车-内饰

JAGUAR C-X17

The concept car Jaguar C-X17 is an icon of the high-end fashion vehicle: it is high-tech, aggressive, and looks to the future.
If the Jaguar C-X17 was set up with the interiors depicted in picture 09 (those of a Rolls Royce Phanton), anyone would realize immediately that there is something wrong: the clear perception of inconsistency would be due to not respecting the principle of coherence.
The principle of coherence would be betrayed by associating the concept car's idea of contemporary design and high tech with some very traditional and antique looking finishes and materials.
The design concept, which is the basis of the design and aesthetics of a car, must be reflected in all its parts, including the interiors.

4 基本原理和法则
法则
理念
4 bases and principles
principles
ideas

4 基本原理

一项好的设计，不管是日常用品设计、建筑设计还是城市设计，都必须有三项基础支撑要素：

- 人
- 框架环境
- 时代性

设计规划也就意味着重视这三大要素互相协调一致的原则。

每一个设计项目，每一个设计成品都必须符合这三项基本原理，即

- 必须满足人们的实际需求，
- 必须与成品所处环境的规范和文化相适宜，
- 必须与成品产出历史时期的技术相关。

4 BASES

The design, whether it is for an object of daily use, a building or a city, is based on three pillars that form the basis, the foundation of a good project:

- man
- framework conditions
- time

Planning means respecting the principle of coherence in relation to the three pillars.

Each project, each realization, must be consistent with the three fundamental principles, namely it

- must respond appropriately to the real needs of man ,
- should be adapted in accordance with the rules and the cultural context in which it is inserted ,
- must relate to the technology available in the historical moment in which it is conceived.

人

人是规划环节中的一个关键因素。

制成品定义的核心就是人们的需求，而人们在使用这件制成品的时候这项需求必须能够得到满足。

设计师所肩负的任务就是通过创造性活动让人们的需求得到满足。

这项独创性的活动是产出成品必不可少的环节，而设计产品是一个复杂的过程，设计师不仅要考虑产品的功能需求，还要考虑社会需求以及那个时代其他方面的多元需求。

11 列奥纳多·达·芬奇，维特鲁威人

MAN

Man is the key element of the planning process.
At the center of the process that leads to the definition of the artifact are the human needs that must be met through the use of the designed artifact.
Meeting human needs is the task that the designer has to attain through a creative act.
The creative act is necessary to conceive a design and the design process is complex because, in addition to mere functional needs, the designer must respond to social needs that are more and more multi-faceted, correctly interpreting the demands and opportunities of his time.

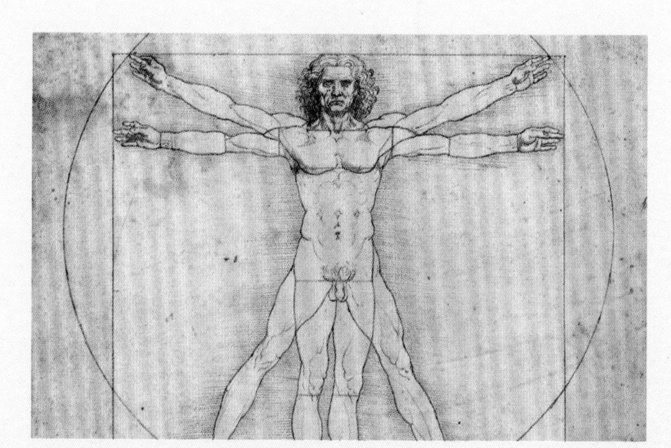

11 - Leonardo da Vinci, Vitruvian Man

12 - cutlery (europe)

13 - chopsticks (china)

框架语境

框架语境是指客观环境和文化环境所设定的元素，对设计项目的理念有着直接的影响。

框架语境这个词看上去会让人觉得像是对设计师思维和行动的一种拘限，但是如果我们用开放的思维和创新的精神去诠释它，那它将成为设计灵感的巨大源泉。

即使是同一项人类需求也会因为时代、文化和社会总体环境的变化而产生出不同的设计。

12 餐具(欧洲)
13 筷子(中国)

FRAMEWORK CONDITIONS

The framework conditions are the elements given by the environment, both physical and cultural, that channel the idea of the project.

At first sight, the framework conditions may be a limit to the thought and action of the designer but, if interpreted correctly with an open mind and creative spirit, they represent a powerful inspirational force.

A mutation of the temporal, cultural or social framework conditions can give rise to different design responses for the same requirement of man.

饮食

人类的首要需求就是“吃”（把食物放进嘴里）。
在欧洲文化和亚洲文化中，解决人类这一需求的方式却截然不同：刀叉和筷子。

EAT

Man's primary need: to eat (bring food to the mouth).
Different answers to the same primary requirement in the European and Asian cultures: cutlery and chopsticks.

时代

这个项目可称得上是“时代的产物“。
一项好的产品应该与它所处历史时代有一致的体现，产品采用当时的技术和材料，因而通过产品能辨识出历史的痕迹。

14 电话(1890)
15 智能手机(2013)

TIME

The project must be "child of his time".
A good product should be a coherent expression of the historical period in which it is conceived, utilizing the available technologies and materials, thus becoming recognizable in time.

14 - phone (1890)

15 - smartphone (2013)

沟通

人类的主要需求：远距离通信。
在不同时代，解决人类远距离通讯的方式不一样：带有拔手选择的曲柄话机，带有触屏的智能手机。

COMMUNICATE

Man's primary need: to communicate at a distance.
Different answers to the same primary need separated by a century: crank phone with number selection by disc and smart phones with touch screen.

4 法则

我们可以从任何优秀的产品、建筑和城市设计中发现优秀设计所遵从的基本原理，也是创造性设计活动中最根本的原理，即

- 形式要符合功能要求
- 简洁
- 可辨识度高
- 尊重能源
- 善于利用现有技术

在运用这些原理时还必须注重内在连贯一致这样一种概念，原理的运用必须相互协调一致，只有这样，设计项目的根本理念才能得到落实和体现。

4 PRINCIPLES

The rational part of the creative process that is the basis of good design is based on the fundamental principles that we find in any good product of design, architecture and urbanism:

- form follows function
- simplicity
- strong and recognizable idea
- respect for energy
- use of the most suitable existing technology

The application of these principles must respect the concept of coherence in the sense that their application should be coordinated in such a way as to allow the consistent affirmation and recognition of the idea on which the project is based.

16

17

18 - ferrari f12 berlinetta

形式追随功能

外在形式，不管是汽车的，还是建筑物的或是一座城市的，都必须追随功能，而且与功能直接相关。

形式是设计成品功能的外在表现。

产品的功能必须能通过产品的外形和产品的部件识别出来。

18 法拉利12 berlinetta

FORM FOLLOWS FUNCTION

The shape, be it that of a car, a building or a city, must flow from its function and be directly related to it.
The form must be a direct expression of the intended function of the designed item.
The function must be recognizable by the shape of the object and by its parts.

法拉利12 berlinetta

汽车的流线型设计就好像汽车被风吹塑而成，在风中疾驰而毫无阻力。

FERRARI F12 BERLINETTA

The sinuous lines of the car evoke the wind that touches the sides of the bodywork.
It seems that the wind has sculpted the car, which is predestined to run fast in the wind.

简洁

KISS法则（keep it simple stupid）在不同的领域（从军事到营销）有着广泛的运用。

这项法则的中心概念就是高品质的设计能让产品尽可能的简单，使用起来直观方便。

简洁是任何创造性设计的策略目标。

SIMPLICITY

The KISS principle (Keep It Simple Stupid) is known and applied in various fields (from the military to marketing) for the elaboration of strategies.
The central concept behind this principle is that high quality design tends to make the product as simple as possible and intuitive to use.
Simplicity should be a strategic goal of any creative enterprise.

19

20

触屏设备

还有什么比触屏产品更简单直接？

给幼儿一个触屏电子产品，他就会本能地靠点触来翻看里面的内容了。

要是运用程序设计得很智能化，不一会儿，幼儿就能在这个新的数字世界里随心所欲地游走啦。他会点开不同的游戏，让它们蹦出各不同的声音和图片，还会给他的设备下不同命令。

TOUCH SCREEN DEVICES

What is more simple and straightforward than a touch screen? Just place it in the hands of a small child and he will begin to explore the hidden content in an intuitive manner.
After a short time, if the applications are designed in an intelligent way, the child is able to move at will in the new digital world, hobnobbing with the most disparate games, making it gush images and sounds, and in the same way giving orders to his house.

注重能源

在当今社会，保护能源和环境的可持续理念越来越重要。

能源，环境，土地都是息息相关的。

我们没有权力浪费资源，污染环境：设计师对整个地球和未来的人们担负着重大的责任。所以今天，职业道德规范和关注能源问题在任何设计领域都是很重要的一个方面。

RESPECT FOR ENERGY

In contemporary society,
the concepts of energy conservation
and environmental sustainability
are becoming increasingly important.
Energy, environment and land use
are directly related themes.
We have no right to waste, to pollute:
the designer has a great responsibility
towards the planet and
future generations.
Ethics and hence the focus on energy
issues today play a key role in design
at any scale.

21

22

设计师应该时刻关注他的创意带来的影响。
创造是消耗能源，利用资源的行为。设计师所肩负的环境责任是巨大的，因为设计师的作品很有可能具有伤害性，会影响未来几代人的社会发展。

The designer cannot forget to reflect
on the impact of what he is creating.
To create is however to act and action
implies energy consumption,
use of resources.
The environmental responsibility of
designers is enormous
and the consequences
of the designer's work can be harmful,
affecting the development of society
for several generations.

技术

设计师应该采用最合适的技术以期适应功能要求，达到设计目标。

设计师要开发新的概念和产品推动社会前进，就必须做到不仅广泛熟知自己业务领域，还必须了解其他生产领域内的现有技术。

23 罗马水道桥(公元前1世纪)
24 阿拉密洛大桥-塞维利亚-西班牙(1992)

TECHNOLOGY

The designer must use the available technology best suited to the function and objectives to be achieved.
A wide knowledge of available technologies both in his own business sector and in other productive sectors is a prerequisite for the designer in order to develop new concepts and products for the advancement of society.

23 - roman aqueduct (I century BC)

24- alamillo bridge - Sevilla - Spain (1992)

2000年前所建的桥都是靠拱形支撑，而不是采用土木工程领域的没有中间物支撑也能远距离架桥的静态技术。从细节来看，大桥还用了大量实物材料。20世纪以来，拉索的运用和结构的改变推进了建造技术，没有中间物支撑也能远距离架桥，而且还节省了大量材料。

2000 years ago bridges and aqueducts were supported by arches because the civil engineering technology of the time (which was based predominantly on compression) was unable to offer efficient solutions for overcoming great distances without intermediate supports.
Therefore, the subdivision into small bays required a massive use of material.
Since the 20th century however, the great advances in construction technology, due to strained tendons and structures, allow us to overcome enormous distances with less intermediate support, resulting in a decrease of the mass.

25 - wasp

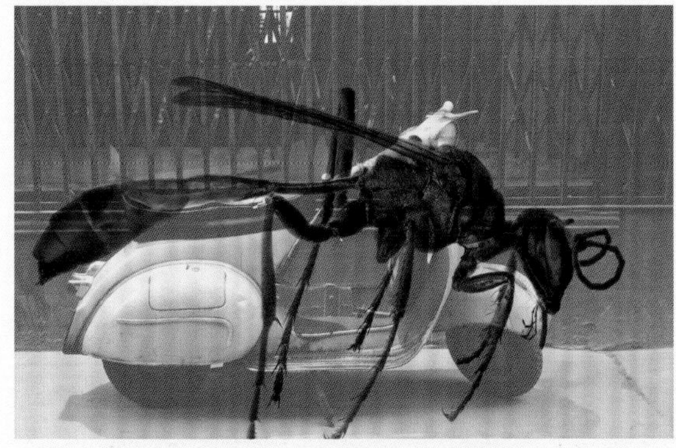

26 - tha animal and the object

27 - vespa 150 gs, piaggio (1955)

理念辩识度高

好产品的一个重要特征就是设计的基本理念和法则极易辨识。

25 黄蜂
26 动物和产品
27 维斯帕150gs比亚乔(1955)

VESPA PIAGGIO

Vespa 是由比亚乔 (Piaggio) 公司制造的一款踏板车，1946年由一名飞机工程师Corradino D'Ascanio设计。车身由冲压钢制成，车体能自行支撑。Vespa 是工业重构过程中的一个重要实例。在“一战”到“二战”期间比亚乔公司主要生产铁路交通工具和飞机。“二战”末期，Corradino D'Ascanio 将飞机外壳的生产转换成其他产品，于是设计了一款新型的摩托车。
Vespa 是意大利经济复苏期一款标志性产品，迄今仍深受人们的喜爱，而且一直被认为是意大利最成功的产品设计之一。Vespa的成功揭示了设计交通工具一条很清晰的理念，那就是简单、低成本、轻便，可以让人们穿着优雅的裙子舒舒服服地去上班。这就是创新。

STRONG AND RECOGNIZABLE IDEA

A good product is characterized by the recognition of the idea and of the principles that were the basis of its design.

VESPA PIAGGIO

The Vespa is a model of scooter manufactured by the Piaggio company from 1946 according to the project of the aeronautical engineer Corradino D'Ascanio. Characterized by a self-supporting body using pressed steel, the Vespa is an important example of industrial restructuring.
Piaggio, in the period between the first and second world wars, produced railway vehicles and airplanes.
At the end of the second world war, Mr. Corradino D'Ascanio diversified production beyond airplane shells by designing a new type of motorcycle.
The Vespa is one of the iconic objects of the Italian economic boom, still loved and considered one of the most successful products of Italian design of all time.
Behind the success of the Vespa, there was a clear idea of a vehicle that would be simple and low cost, a jaunty motorcycle that would allow one to go comfortably to the office with an elegant dress or a skirt.
This is innovation.

4 理念

如果说法则是设计项目的躯干，那么理念就是设计项目的灵魂。
一项高品质设计最强有力的一个理念就是

- 简单

简单，让设计师对于设计的理解和认识更加清晰，是每一项优秀设计的根本所在。

- 创新

创新是设计项目的重要基石，是新产品区别于旧产品的重要特点。

- 可辨识

成品必须能保证设计理念的可辨识度，这样产品才有自身固有的特点，产生巨大的魅力。

当设计师和使用者一看到成品后就能用简单的一个动作、一个符号概括出产品的理念，那么这个设计理念就是非常明显的。在突出理念的基础上开发的产品有着它自己独特的灵魂，从而成为被历史认可的产品。

4 IDEAS

The idea is the soul of the project as the principles are the physical body.
For the idea to be strong enough to be transformed into a winning project, it must be

- simple

Simplicity, the basis of every good design, allows a clear understanding of design intentions and identification of the mind of the designer.

- innovative

Innovation is the cornerstone of the project since it characterizes the product, differentiating it from what already exists.

- recognizable

Recognition of the original design idea must be retained in the finished product, giving the product itself intrinsic strength and great attractiveness.

When both the designer and the user can summarize the idea behind the project in a gesture, a graphic sign or with a simple scheme after simply looking at the finished product, we are facing a strong idea.
The product developed with coherence on the basis of a strong idea has its own recognizable soul and will live in time.

28 - the matterhorn

29 - toblerone (1908)

30 - toblerone (1908)

TOBLERONE
(三角巧克力)

这款巧克力是1908年由Theodor Tobler制造的。Toblerone 和 Matterhorn（马特洪峰）都是瑞士的标志。
这款历经一个世纪，产品设计甚至包装都不曾改变，但销售却常胜不衰，魅力不减的产品其秘密何在呢？也许，秘密就在于它内在的特质 —— 简单。它的造型是很简单的三角形，就像外包装上图示的马特洪峰（国际知名的瑞士标志）。由此，Toblerone也作为一种民族象征悄然兴起。从侧面看，这款巧克力是很多重复的三角形，这种重复让原始设计理念得到更大的加强。从末端掰下一块就是一个小三角。很聪明的设计！

28 马特洪峰
29 三角巧克力(1908)
30 三角巧克力(1908)

TOBLERONE
A product created in 1908 by chocolate maker Theodor Tobler, Toblerone, is, along with the Matterhorn, a symbol of Switzerland.
What is the secret of the success of this product, which for over a century has maintained its charm and is sold virtually unchanged in design and even in the packaging?
Perhaps the secret lies in its unique inherent simplicity: it is a pure shape, a triangle equal to the Matterhorn mountain (one of the best known symbols of Switzerland internationally), which even the packaging refers to.
The Toblerone rises as a national symbol.
Viewed from the side it is a repetition of triangles, which gives even more strength to the original idea.
Just squeeze the bottom of the bar with your finger to tear off a bite, a small triangle.
Brilliant. Delicious.

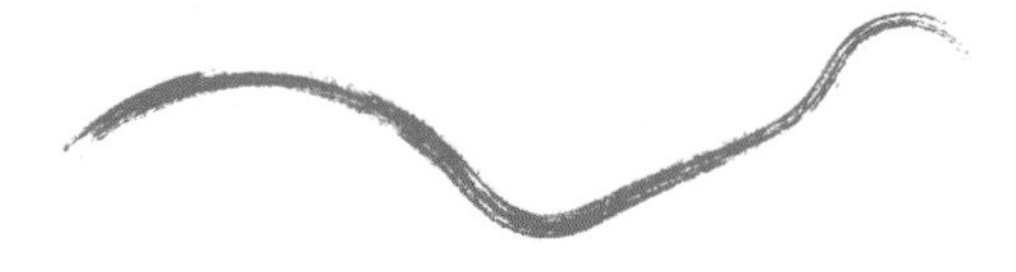

31

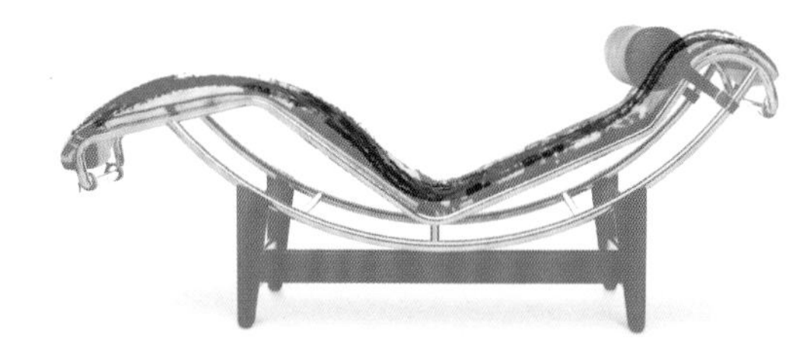

32

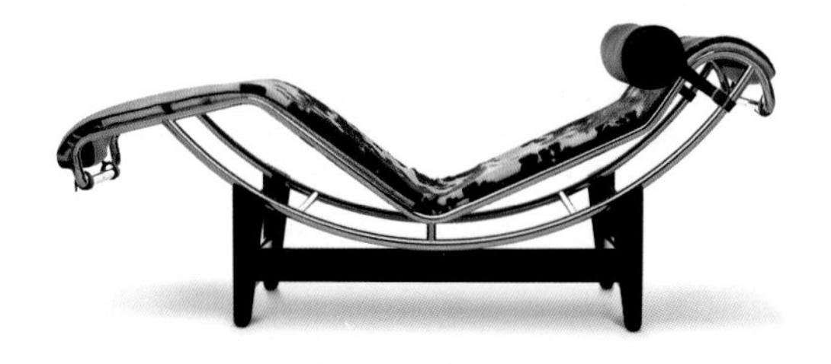

33 - chaise longue LC4 (1929)

躺椅LC4

1928年由Le Corbusier, Charlotte Perriand设计，最初由Thonet生产，1965年由Cassina生产，迄今还在各处有售。它的造型是一段流畅的曲线：椅子看起来像是要接住躺在上面的身体，轻轻地摇着。
线条简洁流畅，产品经久不衰。

33 chaise躺椅LC4(1929)

CHAISE LONGUE LC4

Designed in 1928 by Le Corbusier and Charlotte Perriand, the LC4 chaise longue, initially produced by Thonet and since 1965 produced by Cassina, is still widely sold today.
Its shape is characterized by a smooth curve: the chair seems to take over the body that lies in it and rocks him gently.
Simple, linear, timeless.

34

35

36 - porsche 911 (1964)

保时捷911

保时捷911无疑是最经典，最成功的车型之一。
从设计角度来看，60s可被简化为两条线，一条腰线，一条车顶线，两条线以锐利的角度往下走与车灯和档风玻璃相结合。
从保时捷911的设计还能找到60s当时呈现的独特之美。

36 保时捷911 (1964)

PORSCHE 911
One of the most long-lived and successful cars is undoubtedly the Porsche 911.
At the level of design, the basic idea of the 60s can be reduced laterally to two lines, one at belt level and the other corresponding to the roof, both with a sharp fold down for the headlights and the windshield.
The aesthetes still look back with nostalgia at the unmatched elegance of the original 60s design.

37

38

39 - Aquarama, Riva (1962)

出水丽娃(Riva Aquarama)游艇

出水丽娃快速游艇由工程师Carlo Riva设计，从1962年到1996年都有生产。从开始生产的那一刻起，它就成为了一个真正的传奇，它还享有别的名号，如“游艇中的斯特拉迪瓦斯 (Stradivarius)”、“海上法拉利”，“法国里维埃拉 (French Riviera) 的皇后”。游艇的造型优雅，速度极快，购买游艇的顾客也是名人大腕，因此出水丽娃被认为是全世界最好的游艇之一，迄今还享有传奇美誉。
著名的商贾名流，皇室贵族都以拥有丽娃游艇为荣。其中就有Brigitte Bardot、Anita Ekberg、Elizabeth Taylor和Richard Burton夫妇、Ranieri、约旦国王Hussein George Clooney、Elle Macpherson。
游艇规整优雅的造型让它就像海浪中呼啸而过的子弹，即使在今天，仍是魅力不减。

39 出水丽娃(1962)

RIVA AQUARAMA

The Riva Aquarama speedboat
designed by engineer Carlo Riva
was produced continuously from 1962
to 1996. From the moment it began
to be produced, the Aquarama quickly
became a true legend,
earning extraordinary titles such as:
"the Stradivarius of the boats",
"the Ferrari of the sea" or
"Queen of the French Riviera ".
Thanks to the elegance of the forms,
the speed it reached and not least the
reputation of many of those
who bought these boats,
the Aquarama was considered
one of the best boats in the world
and still retains the aura of legend.
Well known actors, the jet set, royal
families: the famous owners of these
boats were and still are numerous.
Among them we can mention:
Brigitte Bardot, the owner of a Florida;
Anita Ekberg, who preferred a Triton;
the couple Elizabeth Taylor
and Richard Burton who used
one Junior as a tender for their yacht;
King Hussein of Jordan;
the Shah of Persia;
Prince Ranieri and even
George Clooney and Elle Macpherson.
The regular and elegant shape of the
Aquarama resembles a bullet whizzing
through the waves of the sea
and keeps intact its charm even today.

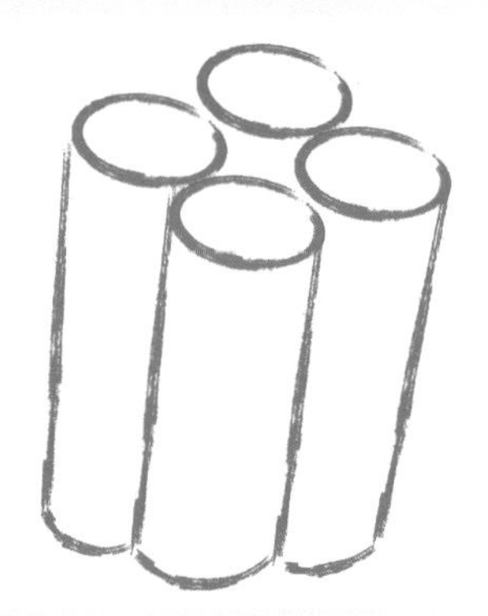

40

41

42 - bmw 4-zylinder, munich (1973)

宝马 (BMW) 4缸大楼

著名汽车制造商宝马的办公总部在巴伐利亚的慕尼黑，1973年开始启用，是德国最好的建筑之一。没有其他的建筑比4缸大楼更能让人将它与汽车工业联系在一起。
大楼的名字"4缸"就解释了为什么大楼有如此外型，它就是参照汽车引擎而建的。
这项设计的理念成功地说明了大楼的主人、用途和它开展的活动。

42 宝马 (BMW) 4缸大楼，慕尼黑 (1973)

BMW 4-ZYLINDER

The tower, which houses the offices of the carmaker BMW in Munich of Bavaria and opened in 1973, is one of the most famous buildings in Germany.
No other building in the German's collective imagination is linked to the automobile industry itself more than the "4-Zylinder".
The name "4-cylinder" clarifies immediately the reason of its shape, establishing a direct reference to the car's engine.
The project idea masterfully expresses the relationship of the building with the activity that generated it, has funded it and uses it.

5 灵感来源
5 sources of inspirations

5 灵感的来源

千百年来，历经各种趋势的影响和文化的浸染，人类的设计活动也变得形式多样，各有千秋。

设计师们一直都在不停地寻觅能帮助他们构想出新创意的灵感源泉。

设计师们经常用各种方式来充盈他们的创造力，要么从以往的设计中获取灵感转化成自己的设计，要么从其他领域发展的可预见未来中获取，要么从大自然中获取，要么从人类活动的各个领域和人类的心理活动元素中获取。

灵感的主要来源：

- 人类和大自然
- 框架语境
- 所处历史时期

5 SOURCES OF INSPIRATIONS

Over the centuries, human design activity has been influenced by various trends and cultural streams that have given rise to different styles.

The designer has always looked for sources of inspiration to help him conceive new ideas.

He who designs often feeds his creative charge by taking a cue from the past, processing and transferring it into his own project, or from the visions of possible futures developed in other areas, or is inspired by nature or elements from various fields of activity and the human psyche.

The main sources of inspiration are :

- man and nature
- framework conditions
- time

灵感 —— 人与自然

人类的需求，感知，精神

人类的需求、感知和精神是设计师们获取灵感和参照的一个重要来源。

人类情感如：

- 感官欲望
- 速度
- 精神

这些人类情感同样是艺术家、设计师们有意识或无意识地汲取深度灵感的重要来源。

INSPIRATION — MAN AND NATURE

HUMAN DRIVES, PERCEPTIONS, SPIRITUALITY

The human drives, perceptions and spirituality are another important source of stimuli and references for the designer.
Feelings like :

- eros
- speed
- spirituality

have always been important emotional and conceptual spheres from which not only artists, but also designers have drawn, consciously or not, deep inspiration.

43 - bocca, gufram (1971)

44 - tatlin, edra (1989)

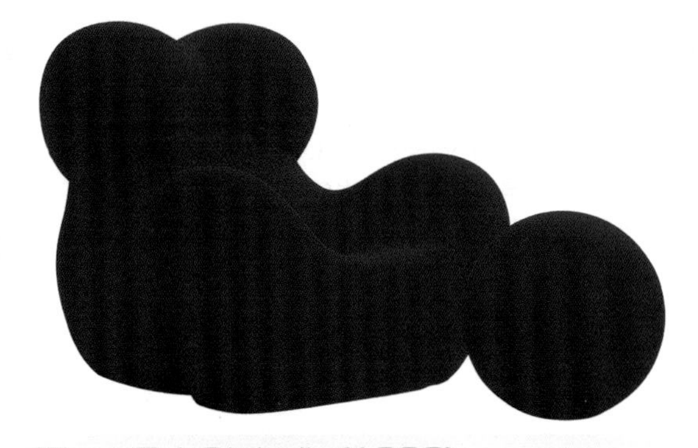

45 - up5, b&b italia (1969)

唇形沙发

能够引起丰富感官意象的造型和颜色就像是引发兴趣的催化剂，因此，广告业经常用它们来吸引公众的眼球，而设计行业则用它们来让产品变得更有吸引力。

43 唇形沙发(1971)
44 塔特林结构沙发(1989)
45 up5沙发(1969)

EROS

Shapes and colors related to the collective erotic imaginary can act as catalysts of interest,
and for this reason they are often used both in advertising to capture the attention of the public and in design to make products more appealing.

意大利设计的座椅

与感官意象相关的造型元素与浓烈的色彩结合能更加吸引用户的注意力。

ITALIAN DESIGN, CHAIRS

The formal elements related to the erotic imaginary combined with the use of an intense red strongly attract the user's attention.

46 - unique circule yacht, Z.Hadid (2013)

47 - galaxy Soho, Z. Hadid (2012)

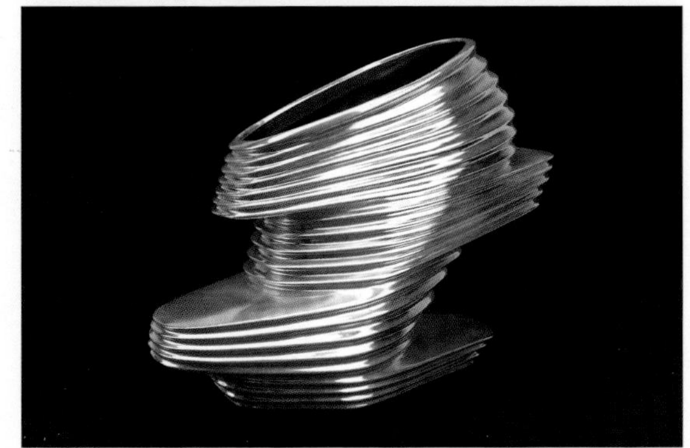
48 - nova shoes, Z. Hadid (2013)

运动、力度、速度

运动、力度和速度不论从客观还是感知角度来讲都是深植与西方大众想象中的概念。在很多国家，汽车无疑成了速度定义的实物载体。而实际上，汽车已经成了崇拜和社会地位的象征。

激起速度的感知有很多种有效方式，如通过重复拉紧以及流线型线条来表现速度。

46 环状结构游艇, Z.Hadid (2013)
47 银河Soho, Z.Hadid (2013)
48 新星鞋, Z.Hadid (2013)

HADID
建筑师Zaha Hadid设计的主要特点就是设计的造型大多是能够传达动态、运动、速度概念的流线造型。
她设计的游艇线条让人联想到海浪的运动，北京Soho区的优雅线条让参观者感觉站在了地平线上。女士新星鞋关注度有升无减，不过鞋的功能和外型还有些不协调。

MOVEMENT, STRENGTH, SPEED

Movement, strength and speed, in physical and perceptual terms, are concepts that are deeply rooted in the Western collective imagination. In many societies, the materialization of speed is by definition the car, which in fact becomes an object of worship and a status symbol. Speed can be evoked in several ways including the repetition of taut and sinuous lines, which appears to be particularly effective.

HADID
The sinuous shapes that characterize the designs of the architect Zaha Hadid transpire dynamism, movement and speed.
Her boat's lines evoke the movement of the waves on the sea, while the elegant lines of the Soho district in Beijing seem to project the viewer toward the horizon.
Instead, the shoe for ladies, Nova Shoe, where function and form seem to collide, raises some concerns.

精神

人的内心都有一种愿望，希望自己能与天相接，升入天堂。
想要创造一个人人敬畏，铭记在心的地方和物品是一项非常艰巨的任务，成功的设计案例凤毛麟角。
天堂就是至高、遥远、完美、永恒的代名词。
因此，设计师需要深挖这种意指天空、象征、简单的文化和灵魂，以光为工具创意无穷。

49 光之教堂，安藤（1989）
50 光之教堂，安藤（1989）

SPIRITUALITY

Inherent in man is the desire
to connect to the sky,
to rise to the divine.
Creating objects and places of worship
and memory is a difficult task
that often ends in failure.
The divine is in his definition high,
distant, perfect, eternal.
Hence the need to dig deep
in the culture and in the soul aiming
at the sky, at symbols and simplicity,
working on light as a vehicle towards
the infinite.

49 - church of light, T. Ando (1989)

50 - church of light, T. Ando (1989)

安藤（ANDO）
安藤设计的这所地处茨城的教堂后墙上切开了一个十字形，通过光来呈现出一个神圣的标志是这项设计的精髓和最终目的。

ANDO
With the introduction of a cut
in the shape of a cross on the back wall
of his church in Ibaraki, the architect
Tadao Ando generates, through light,
the sacred symbol that becomes the
purpose and essence of the project.

灵感的源泉 —— 大自然

无论过去还是现在，技术、设计、规划领域通常都将大自然视为灵感的——重要来源。
大自然的方方面面及各种现象都给了人们无尽的设计灵感。

51 耳机，温为才教授

NATURE AS A SOURCE OF INSPIRATION

In the past as well as nowadays, technology, design and planning in general, have often looked to nature as one of the main sources of inspiration.
Not only the forms but every aspect of nature and its phenomena have offered and still offer the viewer an endless amount of ideas and solutions for innovation and design.

51 - earphones, prof. Stone Wen

耳机

中国设计师温为才（Stone Wen）设计的耳机，其灵感就直接来自于大自然。整副耳机的外廓就像一朵美丽的郁金香。

EARPHONES

The design of the earphones developed by Chinese designer Stone Wen is inspired directly from nature.
The union of the two earphones reveals the silhouette of a beautiful tulip.

源自框架的灵感

设计师还能利用来自产品图形学的技巧，更好地表现和传达设计理念。没有灵感的时候可以试试这种方法。

一些可以突出或弱化产品或产品某个部件的技巧：

- 光线运用
- 色彩运用
- 重复
- 复制模仿

INSPIRATIONS - FRAMEWORK

The designer in his work can take advantage of the use of techniques derived mostly from graphics that allow him to express and convey the idea in the best possible way; the same techniques can be used in case of a lack of ideas.

Some of the techniques that allow the designer to highlight or overshadow the product or parts of it are, for example

- use of light
- use of color
- repetition
- copying

光

通过对光的调整（明暗、光影的调整）可以给物品或环境增添另一种意境。一般情况下，亮一点的色调会达到一种静谧的效果，而暗黑色调会营造出一种戏剧性效果。

52 美军公墓
53 德军公墓

LIGHT

Through modulation of the light (management of tones and shades), it is possible to load the object or the environment with additional meaning. Generally, an effect of serenity is transmitted by light tones and a dramatic effect by dark tones.

52 - American military cemetery, colleville-sur-mer

53 - German military cemetery, pargny-filain

纪念诺曼底登陆，战争公墓

法国的战争公墓埋葬着“二战”时期诺曼底登陆盟军阵亡士兵的遗骸，到访墓地的人们都会被深深地触动。
尽管两个墓园埋葬的都是为自己国家英勇作战的年轻士兵，但是两个墓园给到访者的感觉却大不相同。在盟军的墓园，十字架是白色的，造型瘦长，传递的是一种平和、安静、公正的意象；德国墓园的十字架是黑色的，造型粗壮，给人一种负罪感。

CEMETERIES OF WAR, NORMANDY

The war cemeteries in France that house the remains of soldiers killed during the Allied landing at the end of World War II, leave a deep emotion with visitors.
Despite the fact that in both cemeteries thousands of 20 year old young men have been buried, who each were fighting for their own nation, the feeling in the Allied and German cemeteries is diametrically opposed.
The Allied cemeteries, thanks to the white color and slender shape of the crosses, in general convey an idea of peace, serenity and justice;
German war cemeteries, because of the dark color of the crosses and their often squat and heavy shape, convey an overwhelming sense of guilt.

54 - red kilometer, J·Nouvel -

色彩

颜色的运用需格外谨慎。
只有一种色调呈现而看不到其他色彩的时候，单色调的表现力度最大。
在建筑物上运用单色调的方案更令人信服：在一个中性色调（白色-灰色-黑色）的环境下放入一个亮色的建筑会将颜色的影响发挥到极致。

COLOUR

The color should be dosed carefully.
The strength of the single color is expressed at its highest when there are no other visible colors.
In architecture, monochromatic solutions provide more convincing results: enter a colored object in a neutral (white - gray - black) environment and the object will give the best of itself.

55 - red kilometer, J·Nouvel

红色公里墙

能让经过米兰和布雷西亚交界的汽车快速道的人深有印象的建筑恐怕就只有“红色公里墙”了。这是一道与快速道相平行的1000米长的红色墙体。这道墙是一个研发中心大楼的外墙。
简单的创意、线条、色调，但是这项设计却非常成功。

54 红色公里墙，让·努维尔
55 红色公里墙，让·努维尔
56 wozoco,老年公寓－阿姆斯特丹

RED KILOMETER

The only building that remains imprinted on anyone passing on the Milan - Brescia motorway is the “Kilometro rosso”:
a 1000 meter long red wall that runs parallel to the highway and houses a center for research and development.
A simple idea, linear and monochromatic:
success is guaranteed.

56 - wozoco, mvrdv - Amsterdam

老年公寓

多种色彩的组合运用让人们更关注色彩效果而不是建筑本身。这种方法可能更多地属于美术范畴，跟建筑范畴相距甚远。因此，如果是为了让色彩组合达到最佳效果，大家应该参照美术标准，而非设计方面的法则。

WOZOCO

The combination of very different colors shifts attention from the object to the resulting chromatic effects; this approach is typical of the artistic sphere, and moves away from the architectural one.
To obtain good results with the coupling of different colors it is therefore a must to refer to artistic criteria, not to design principles.

重复

设计师采用重复的手法让表达的概念更强烈，清晰。

57 大屠杀纪念馆，柏林 (1998)
58 大屠杀纪念馆，柏林 (1998)

REPETITION

Repetition is used by the designer to give more strength and clarity to the concept.

57 - holocaust memorial, Berlin (1998)

58 - holocaust memorial, Berlin (1998)

大屠杀纪念馆
柏林大屠杀纪念馆的设计采用了重复简单而厚重的水泥块，每个水泥块的大小又稍有不同，使得整个设计表现出强大的冲击力。
成百上千的水泥块整齐摆放，给到访者一种巨大的情感震撼，也让设计师想要表达的理念清晰可见。

HOLOCAUST MEMORIAL
The enormous force that is transmitted by the Holocaust memorial in Berlin is due to the use of simple and massive blocks of concrete, but above all by the repetition with slight modulations. The hundreds of blocks laid in regular rows convey a strong emotion to the visitor and make the idea that the designer wants to convey immediately clear and almost palpable.

复制模仿

复制偶尔也是允许的，这可不是侵犯版权的违法行为。
人们经常复制一些造型或图像来表达一种概念或是阐述某个内容。
复制手法可以成为一种谴责的方式。
没有创意的时候，建筑师们通常会采用复制仿建这种技巧。

58 威尼斯，意大利 (XV secolo)
59 威尼斯人酒店，拉斯维加斯(1999)

COPY

Copying is allowed: on certain occasions, it is not a crime.
Often shapes and images are copied to express a concept
or to make content explicit.
Copying can be a critical way
to denounce.
Copying without any interpretation is a technique sometimes used
in architecture, usually where ideas are few or completely lacking.

59 - Venice, Italy (XV secolo)

60 - the venetian resort, Las Vegas (1999)

威尼斯

威尼斯：真实的威尼斯城与拉斯维加斯的仿建威尼斯。

VENICE

Venice: the real one and its copy in Las Vegas.

61 - porsche pavillon - Wolfsburg (2012)

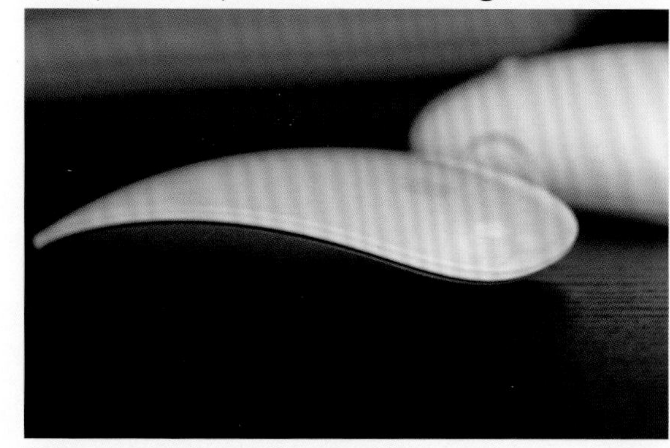
62 - magic mouse Apple

保时捷大厅

最近在沃尔夫斯堡主题公园兴建的保时捷大厅就是从保时捷产品造型上获取的灵感。它的流动型线条和造型让这座大厅看起来更像是一件雕塑品，而不是一幢建筑。
设计师采用了类似于在汽车和航空制造业运用的一项轻结构单体框架技术来建成这样一个外部造型。
这座大厅的外表是由620块不锈钢片焊接起来现场铺装完成的。
这座建筑的整体造型一下就会让人联想到保时捷品牌的经典像征，因而也算是达到了设计的一个初衷。
但不仅仅是这样。如果我们再仔细看看这座大厅的轮廓和外形，我们会发现它与著名的苹果鼠标是一模一样的。

60 保时捷大厅－沃尔夫斯堡 (2012)
61 苹果鼠标

PORSCHE HALL

The Porsche pavilion, recently built inside a theme park in Wolfsburg, draws inspiration not only from the world of Porsche, but goes beyond. With its dynamic lines and silhouette, it looks more like a sculptural object rather than a building,
To build the outer casing, a technology has been used similar to that for constructing the monocoque frames for lightweight structures in the automotive and aerospace industries.
The casing was made with a total of 620 sheets of stainless steel welded together and assembled on site.
The pavilion as a whole certainly evokes the formal characteristics typical of the Porsche brand and, therefore, fully carries out its task.
But not only. If we look more carefully at the profile and lines of the whole, we cannot fail to note that the building has the exact form of a well-known Apple mouse.

源于时间的灵感

过去

我们现在所做的每件事都有一个渊源，来源于过去。

设计经常要参照以前的产品。以前构思设计并制造出来的产品是设计灵感的不竭源泉。重新利用旧的设计可以通过以下方式做一些过滤：

- 再演绎
- 再利用
- 再激活

INSPIRATION - TIME

THE PAST

Everything we do has its origins in the past and is based on the past.
The past is a major source of reference for the designer.
The huge reservoir of what has already been conceived, designed and manufactured is an inexhaustible source of stimuli, ideas and solutions to be reused in the design by a process of filtering through

- reinterpretation
- reuse
- revival.

再演绎

再演绎指的是用一种现代的方式来重新设计以前的产品。现代的方式可以是运用新的材料和技术，或是采用更符合现代品味的外表和造型。运用现代技术创新一个旧的产品,但不影响产品的象征特点和主要功能。

63 菲亚特500 (1957)
64 菲亚特500 (2007)

REINTERPRETING

Reinterpreting is to redesign products of the past in a modern way using new materials and technologies, adapt forms and surfaces to the contemporary taste and fashion using modern technologies, or create a different object from the original while recalling symbols and functions.

63 - fiat 500 (1957)

64 - fiat 500 (2007)

菲亚特500

著名的汽车制造商菲亚特，曾经一度也没有好的创意，公司决定重新使用过去的神话产品 —— 菲亚特500，并对其进行重新设计。菲亚特500是意大利最好的汽车之一，1957 — 1975年间，这款车的销售量是4250000辆。
对于注重设计的设计师来说，这项决定的可行性值得怀疑。但是从市场反应来看这项重新设计的决定是成功的，新款菲亚特500成了欧洲市场的热销产品。

FIAT 500

The carmaker Fiat, in the middle of a crisis of ideas, decided to dust off a myth of the past: after almost 50 years Fiat undertakes the restyling of the glorious Fiat 500, "the italian car" par excellence, in production from 1957 to 1975, with 4'250'000 pieces sold.
The operation is questionable perhaps for design purists, but very appropriate in terms of trade: people like the new 500, which quickly becomes one of the best selling cars in its segment in Europe.

再利用

再利用，包括恢复和重新启用将被销毁的整个产品或某个部件，通过某些改变和处理，让这些产品或部件变成具有其他功用的产品。
通过这种方式让产品得到了新生，而且与原来的功用没有任何关联。

65 废弃的集装箱
66 Freitag总部，苏黎世

REUSE

Reusing consists of the recovery and reuse of whole objects (or some of their parts) that would otherwise be destined for disposal through a process of change and adaptation to a new function.
The new life of the objects recovered in this way generally does not have anything to do with their previous function.

65 - container to be disposed

66 - freitag hq, zurich

Freytag 办公室

Freytag是一家著名的用卡车车篷帆布做环保袋包的公司。位于苏黎世的总部办公室和陈列室是由废旧集装箱堆砌建成的。
公司的陈列室更是公司标志和公司建筑的最佳展示：产品的基本理念和空间理念完美呼应，互相强化。

OFFICES FREYTAG

The office and showroom building at the Zurich headquarters of Freytag, a company famous for its bags produced through the recovery of used truck tarpaulins for the transport of goods, were constructed by stacking a series of used containers to be scrapped because they were no longer fit for transport of goods by road.
The showroom is a wonderful example of coherence between Corporate Identity and Corporate Architecture: the basic philosophy of the product and the space coincide perfectly, reinforcing each other.

再激活

再激活的意思是改变现有日常用品的美学外观，也可以更换使用材料。再激活设计可能出于不同的创作动机，也有不同的意义，可能是出于改善的目的，或是为了适应不断变化的社会环境，也可能是出于某种讽刺目的。

67 潘顿椅，V·Panton (1959)
68 农夫草编椅

REVIVE

To revive is to propose changing the aesthetic look of existing design objects or everyday items, possibly replacing the constituent materials.
Reviving may be impregnated with different motivations and may have many meanings, from the search for improvement to adaptation to the changing conditions of society, or to irreverence.

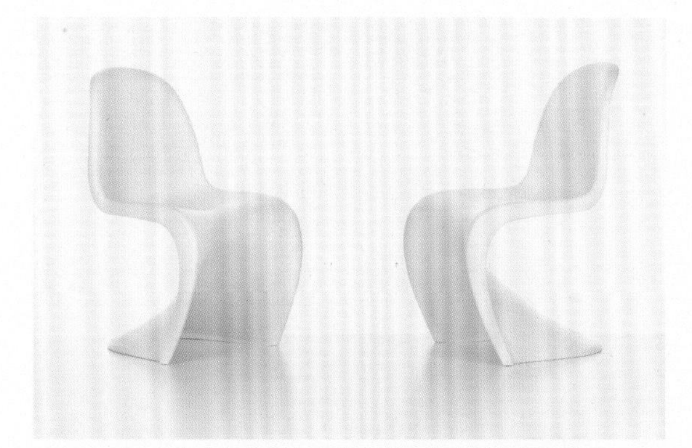

67 - panton chair, V·Panton (1959)

68 - peasant straw chair

Panton 椅

从典型的地中海欧洲乡村座椅到 V·Panton设计的座椅，这两款座椅的设计风格有着星际般的距离。Panton 热衷于采用塑料这种潜力材料，1959年设计出了这款抢眼又有创意的座椅。利用新的材用和新的技术，Panton 重新激起了人们舒服落坐的需求，这款坐椅也在全世界畅销不衰。

PANTON CHAIR

From the chair of the typical rural countryside of Mediterranean Europe to the chair by Verner Panton is a stellar distance.
Panton, fascinated by the potential of plastics, created this fascinating and eye-catching form in 1959.
On the basis of a new material and the related new technologies, he revives the man's need to sit comfortably at the table or with friends, giving rise to a classic of design still sold today throughout the world.

未来

设计师创造未来：未来是由设计师的创意和眼界打造的。

过去的历史是设计师灵感的重要来源，未来的想象同样也是。
设计师常常会预想未来，为更好的未来生活而设计。

其他领域形成的未来构想也能激发设计师的灵感，让设计师从自己的经验局限跳脱出来获取更多的灵感和想象。

电影世界是向大众们展示未来前景的一个窗口，而未来前景是阐释当今人们精神、忧虑和希望的一种方式。
科幻电影最容易让人们理解未来可能是什么样子，设计师也会从这类电影中寻找灵感。

通过看电影的方式，设计师也能构想出自己对未来的想象。

THE FUTURE

Whoever designs creates the future: the future is shaped by the ideas and visions of the designer.

As with the past, the future is also an important source of inspiration for the designer.
Often, the designer tries to predict the future or to make a contribution to bring a better future, designing forms and objects for the man of today and tomorrow.

To go beyond his own experience and his own imagination, thus getting a wider range of possibilities, the designer can be inspired by the visions of the future developed in other fields.

One area that is accessible to the public and yet very challenging with regard to the vision of the future is the world of cinema, which interprets the spirit, fears and hopes of society.
Fiction movies are one of the most convenient ways to quickly get an overview of possible futures, which can be used by the designer as a source of inspiration.

From the cinema, the designer can get ideas to develop his own vision of the future.

69 - the bike of light

70 - tron : legacy (2010)

71 - BMW i8 (2014)

宝马-创：战纪

电影《创：战纪》里那些酷炫的摩托车就是摩托车发烧友的梦想：骑上这样的车，达到人车合一的境界。影片最大的亮点就是赛车的速度与激情，塑料与色彩营造的一种酷炫氛围不禁让人的思绪飞到充满速度的未来。
几年后宝马公司推出了一款新概念车，设计灵感就来自于“创式”摩托车，该车的宣传片简直就是《创：战纪》的克隆版。
这是直白明显的灵感源泉。

69 光电摩托
70 创：战纪 (2010)
71 宝马 i8 (2014)

BMW - TRON: LEGACY

In the movie “Tron: Legacy” bikes are fantastic objects that capture the dream of every sports bike enthusiast: to merge with his bike.
The action is characterized by speed and passion for racing; the fantastic atmosphere created is perfect with plastic and color used to convey emotions linked to the future, racing and speed.
A few years later, BMW introduces its new concept car, clearly inspired by the “tronian” bikes and feelings in a promotional video that looks like a clone of the movie, even in the colors. Undisguised inspiration.

6 相关概念
6 related concepts

6 相关概念

在设计过程中有很多直接或间接相关的概念。

这些概念影响着设计过程。设计师的兴趣和敏感度不同，影响的方式也不尽相同：

- 对美的寻求
- 传承不朽
- 艺术
- 时尚
- 文化

虽然这些概念不属于实际设计过程，但设计师有意无意间都会参照这些概念并受其影响（艺术和时尚）。

有时候设计师会从其他领域的思想中借用一些标志符号，目标物，理想标准，从而将自己的设计推向一个更广阔的空间（美，传承不朽）。

6 RELATED CONCEPTS

Within the design process fit a number of concepts
that are directly or indirectly related to it.

These concepts are elements that interact
with the design process in different phases and ways
depending on the interest and sensitivity of the designer:

- search for beauty
- immortality
- art
- fashion
- culture

Although these concepts are alien to the actual
design process, the designer, whether conscious or not,
refers to it and is affected (art and fashion).

Sometimes borrowing symbols, goals and ideals
from these areas of thought, the designer pushes his work
beyond the limits of the existing, the real,
to a project of greater scope (beauty, immortality).

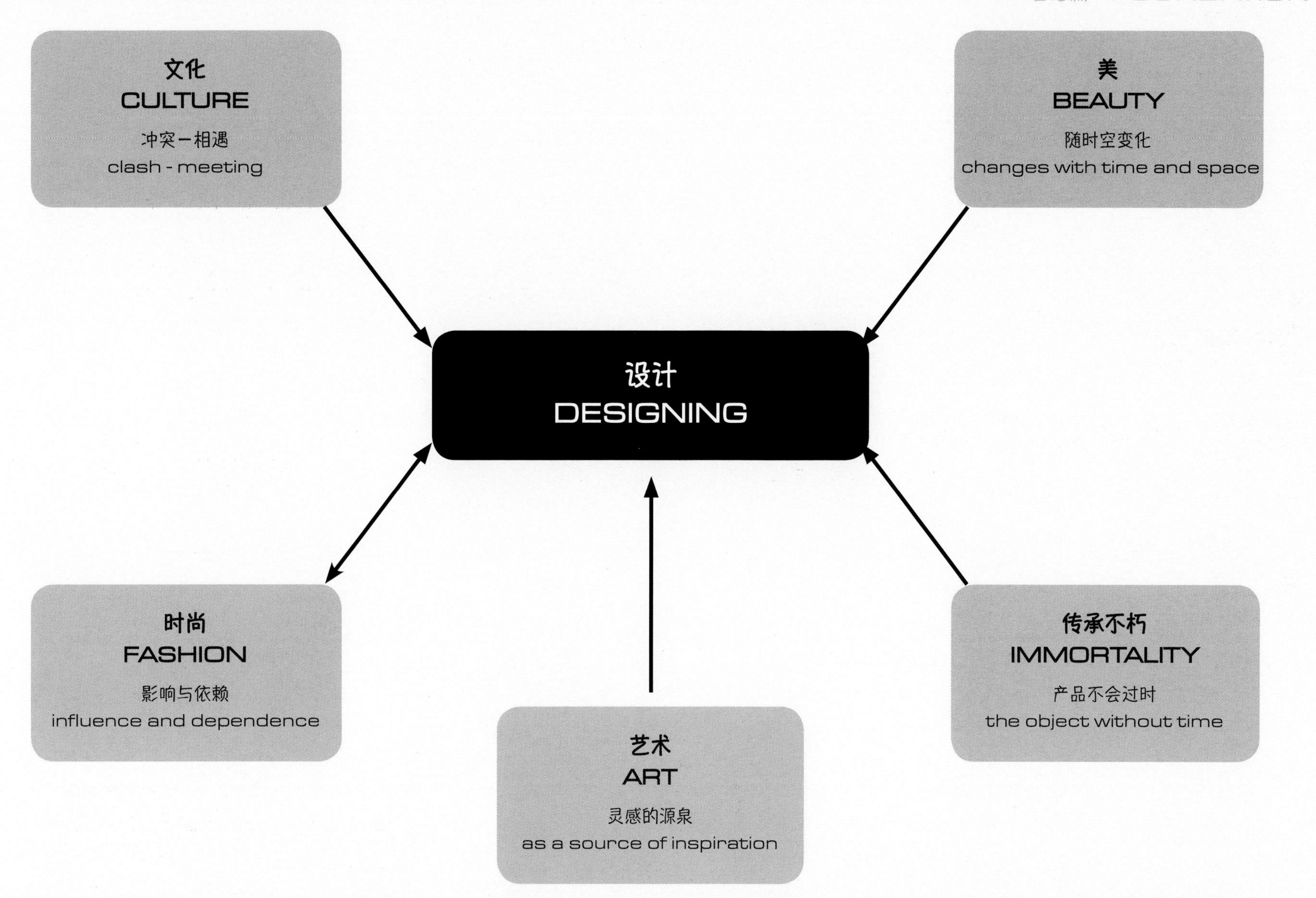
文化
CULTURE
冲突一相遇
clash - meeting
美
BEAUTY
随时空变化
changes with time and space
设计
DESIGNING
时尚
FASHION
影响与依赖
influence and dependence
艺术
ART
灵感的源泉
as a source of inspiration
传承不朽
IMMORTALITY
产品不会过时
the object without time

美

一直以来，审美的标准都随着历史、文化、政治、宗教环境的变化而变化。

例如，在我们已发现的古物资料中，女性的肚子和乳房丰满圆润是生育力强、富足繁荣的象征。这种审美与宗教、神话紧密相关。

历经几百年的变迁，审美的标准已逐渐脱离了宗教意义（至少在西方社会是如此），而且还会继续有所变化。
例如，几百年前红润丰满的女性形象与当今社会的女性美的标准就相去甚远了。

BEAUTY

Since ancient time, aesthetic parameters have changed based on variations in the historical, cultural, political and religious conditions.

Among the representations of deities arrived to us from antiquity we find, for example, female figures with belly and breasts greatly developed, images which refer to the concepts of fertility, abundance and prosperity. The aesthetic was closely related to religion, myth and often magic.

Through the centuries, at least in the West, the aesthetic criteria have gradually been freed from magical-religious significance, but also have continued to mutate. As an example, the florid and even fat appearance of the female figures in the pictorial representations of a few centuries ago is very far from the canons of feminine beauty in contemporary society.

72 - venus of willendorf, ca 30000 a.C

73 - venus with love and a satyr, Rubens

74 - model

沃尔道夫的维纳斯

沃尔道夫的维纳斯是一尊石灰石女性雕像（高11厘米），约于公元前30000年前雕刻而成，现收藏于维也纳自然历史博物馆。
沃尔道夫的维纳斯这尊石雕表明原始艺术中女性审美的一个重要特点就是忽略五官，突出人物的乳房和肚子。
这些特点是旺盛生殖力和富足的象征。

爱神维纳斯和山林之神萨堤

Peter Rubens1614年作，油画，142x184 cm，安特卫普，比利时皇家艺术博物馆
Rubens画笔下的维纳斯，皮肤光润，身形丰满浑圆，充分表达肉欲感官美，与古典艺术的平衡审美截然不同。而另一方面，我们看到走秀的模特都是身形苗条、线条挺直、果敢冰冷，这是西方现代社会审美标准的表现。

72 沃尔道夫的维纳斯，约公元前30000年前
73 爱神维纳斯和山林之神萨堤
74 模特儿

VENUS OF WILLENDORF
Female statuette in calcarea stone known as the Venus of Willendorf (height 11 cm), ca 30000 BC, Naturhistorisches Museum, Vienna.
As in the case of the Venus of Willendorf, the primitive art feminine statuettes often have an aesthetic characterized by a lack of interest for the features and by a considerable enhancement of the breast and of the sides of the figures; this has led us to generally consider them fertility and abundance symbols.

VENUS WITH LOVE AND A SATYR
oil on canvas, 142x184 cm, Peter Rubens, 1614, Antwerp, Koninklijk Museum
The Venus of Rubens, with her glowing skin, rounded and almost rotund in shape, shows an aesthetic now far from the equilibrium of classical arts, that refers to concepts of sensuality and eroticism.

The slender figures, thin and hard at the same time, that we see walking decided and icy on the prêt-à-porter catwalks, could well represent the aesthetic canons of contemporary Western society.

传承不朽

设计师的梦想就是能设计一些经典永恒、传承不朽的作品。

一项超越时间束缚的成功设计秘密是什么呢？

经典永恒的产品是用精艺的方式演绎时代的精神，传达完美意境。

这些产品大都线条干净利落，更重要的是，它们的内在雅致从无定式。

IMMORTALITY

The dream of the designer is to create things
that endure over time, eternal products.

What is the magic formula, the secret ingredient,
needed to transform the project
into a timeless object of worship?

The objects that are long-lasting are those
which interpret in a masterly manner
the spirit of the time
and convey emotions and perfection.

They are mostly characterized by clean lines,
by the essential,
by the intrinsic elegance that never sets.

75 - barcelona, Mies van der Rohe (1929)

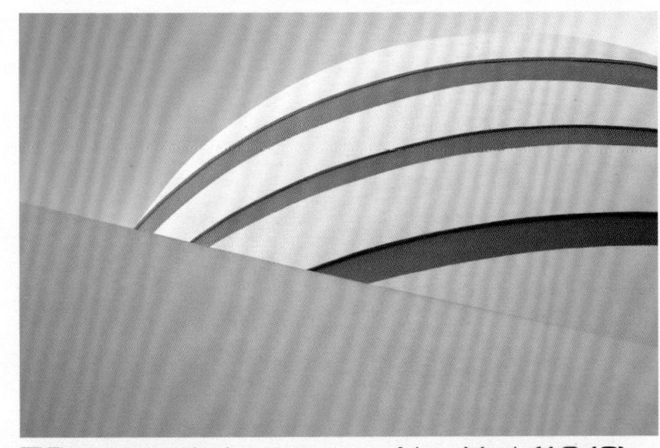

76 - guggenheim museum, New York (1943)

77 - jaguar E-Type (1961)

巴塞罗那座椅
即使是在今天，在很多有名的住所，奢侈品店，很多总部办公室和私人银行的休息室都能看到巴塞罗那座椅。这款座椅是由建筑师Ludwig Mies van der Rohe设计的。他还负责设计了1929 年巴塞罗那世界博览会的德国厅。
这款座椅线条简单，用色沉静，优雅致极，直到今天还在生产。

古尔汗姆博物馆
纽约的古尔汗姆博物馆由建筑师Frank Lloyd Wright 设计，尽管这座建筑已经落成70多年了，但它还是全世界最优雅的建筑之一：它就像是昨天建成的大楼一样。

捷豹 E-TYPE
捷豹 E-TYPE可以说是汽车设计史上最漂亮的车型之一，到现在还是很多汽车爱好者的梦想。
简洁而流动的造型堪称完美。

75 巴塞罗纳座椅, Mies van der Rohe (1929)
76 沃尔汗姆博物馆, 纽约 (1943)
77 捷豹 E-Type (1961)

BARCELONA
Even today in many prestigious residences, in luxury shops, in many representative offices and in the waiting rooms of private banks, abound Barcelona chairs, the chair designed by architect Ludwig Mies van der Rohe for the German Pavilion at the Universal Exhibition of Barcelona in 1929. Simple lines, maximum elegance and sobriety of colors, it is still produced today according to the original design of its creator.

GUGGENHEIM MUSEUM
At 70 years after its inauguration, the Guggenheim Museum in New York by architect Frank Lloyd Wright is still one of the most elegant buildings in the world: it could have been built yesterday.

JAGUAR E-TYPE
One of the most beautiful cars ever designed, the Jaguar E-Type, is still the dream of many car enthusiasts.
The clean and streamlined shape is perfect.

艺术

在当今时代，艺术是设计师灵感的重要来源。

事实上艺术家先是发现深藏于社会生活中的内在本质，然后再用他的作品来表达这种内在本质。这一点可以作为时尚界和设计界的参考。

当今世界的现代艺术已经沦为了地位的象征，整个社会充斥着浓重的商业气息。

幸运的是在这个精神荒芜、灵感枯竭的现状中有些艺术家的作品依旧能够从不同的角度给设计师带来灵感：素材图形，或是感知，或是概念。

ART

In the modern era, art was a primary source of inspiration for the designer.

The artist, in fact, caught the first instincts latent in society and expressed them in his works, which became one of the possible points of reference for fashion and design.

In the contemporary world, art has been reduced to a status symbol; it has become a market governed by purely commercial considerations.

Fortunately in this landscape, so often unedifying for the spirit and devoid of ideas, episodes of artistic genius are not rare and some works of art can provide interesting stimuli to the designers from different points of view: graphic, material, sensory or conceptual.

78 - wave, R. Serra (2004)

79 - the bean, A. Kapoor, Chicago (2004)

80 - balloon dog, Jeff Koons (1994)

艺术和建筑
Kapoor和Serra的雕塑，很有意思的观感体验。

5'500万美元的巨型金属狗
杰夫·昆斯（Jeff Koons）的巨型金属气球狗的拍卖售价达5'500万美元，艺术家Damien Hirst将防腐处理的填充鲨鱼放置在一个玻璃箱里，以1'200万美元拍卖。这些作品向人们展示一种永生的理念，这些都是当今艺术圈流行机制下的产物。

只流俗于时尚界的艺术家们，他们的作品对社会毫无益处，表达的东西也毫无新意。

78 波浪,R. Serra (2004)
79 豌豆,A. Kapoor ,芝加哥 (2004)
80 气球金属狗, 杰夫·昆斯 (1994)

ART AND ARCHITECTURE
The sculptures of Kapoor and Serra, exciting sensory stimuli.

THE 55 MILLION $ DOG
The quasi-inflatable steel giant dog by Jeff Koons, sold at auction for 55 million USD, and the embalmed shark added to a glass case by the artist Damien Hirst (The physical impossibility of death in the mind of someone living), sold at auction for 12 million USD, are some of the best known examples of the mechanism that currently rules the art world.
Works that have no value to society and that do not express absolutely anything new, by so-called artists who have no other merit than that of having inserted themselves into the appropriate circuits...

时尚

设计、建筑、城市规划影响着时尚，同时自身也成为时尚的一部分。
我们看到的时尚是一些能持续几年的暂时现象。

FASHION

Design, architecture and urban planning are strongly influenced by fashion and often contribute to making fashion itself.

We are witnessing some temporary phenomena that can last for several years.

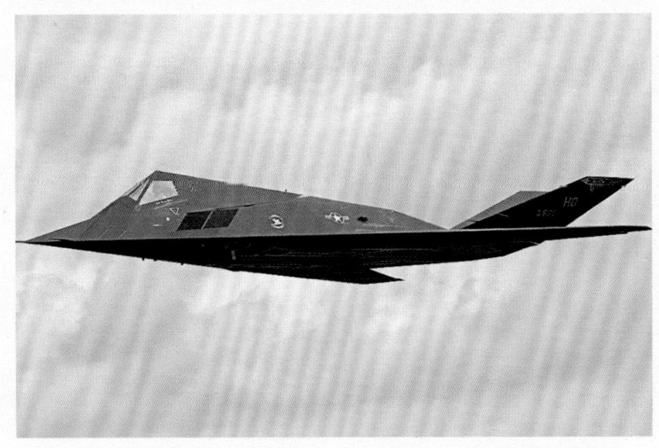

81 - f117 nighthawk, lockheed (1981)

82 - tb group knives (2013)

83 - lamborghini reventòn (2014)

隐形技术

1988年夜鹰F117轰炸机向大众开放展览，隐形技术开始走入寻常百姓家。
隐形物体的表面由一系列切割面构成，表面涂有特殊涂料，传统的雷达系统几乎监测不到它们。几年后这种切割面造型的新技术走进了大众意象：电影院的蝙蝠侠，喷涂亚光漆的超级跑车，还有日常生活中用到的物品。

81 F117 夜鹰轰炸机,洛克希德公司 (1981)
82 tb刀具 (2013)
83 兰博基尼reventòn (2014)

STEALTH TECHNOLOGY

In 1988, the F117 Nighthawk bomber was presented to the public: stealth technology came into our homes.
Characterized by a series of sharp surfaces and specialty paints, stealth objects are (almost) invisible to traditional radar systems.
Some years later, the angular shapes of the new technology gets in the collective imagination: from the cinema with Batman to super cars with matt black bodywork, to objects of daily use.

文化

每一位规划者和设计师都会很自然地参照自己的文化。
另外，设计师会有意识地选择一些自己文化中的个性元素将其汇入到一种不同的文化环境中并加以放大。
设计师总是不断地寻找新的形式和新奇点，他们也可以跳脱自己的经验从异域文化中继续寻找。

84 阿拉伯清真寺

CULTURE

Referring to one's own culture is unconscious and automatic for every planner and designer.
In addition, it is possible to consciously select individual elements of the culture, amplifying the scope and introducing them into an alien context.
The designer, always looking for new forms and new stimuli, can draw from foreign cultures and far from his own experience.

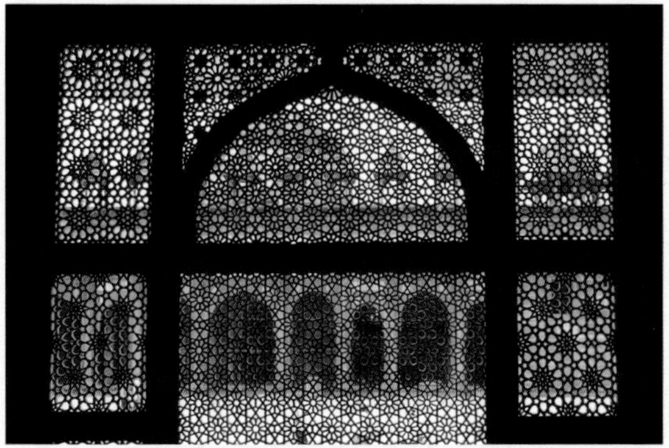

84 - Arab mosque

85 - Arab world institute, Jean Nouvel

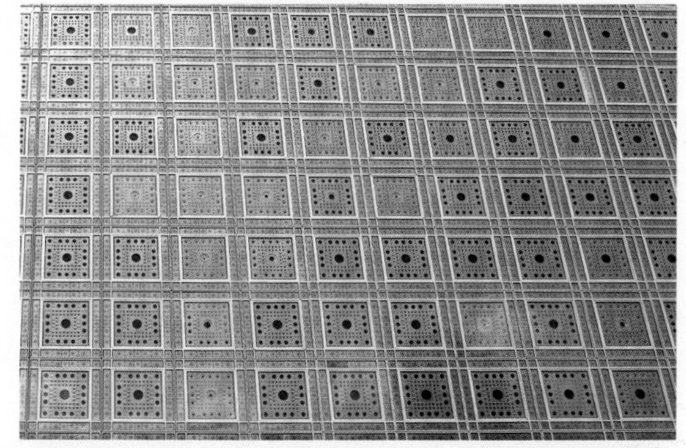

86 - Arab world institute, Jean Nouvel

87 - diaphragm in the facade

阿拉伯世界文化中心

让·努维尔 (Jean Nouvel) 是法国巴黎阿拉伯世界文化中心的设计师。
他将阿拉伯文化与现代化的建筑形象和技术巧妙地融为一体。
这座建筑外墙由形式简单和面积等量的玻璃和钢铁构成，利用光与影的结合变化充分表达阿拉伯文化理念。
外墙的整套遮光系统可以根据太阳光的强弱变化自动调节光影量。努维尔设计的外墙很容易激发人们对几何学这种典型阿拉伯文化的联想，感受伊题兰文化象征符号的奇特魅力。

85 阿拉伯世界文化中心, 让·努维尔
86 阿拉伯世界文化中心, 让·努维尔
87 正面光影图

INSTITUTE OF THE ARAB WORLD

Jean Nouvel, finding himself confronted with the task of designing the Arab World Institute in Paris, managed to masterfully combine the idea of Arab culture with a modern image and technological progress. The architect designed a building with simple and regular volumes of glass and steel giving the task to light and shadows to convey the idea of the Arab world. Through a series of baffles that automatically move according to the incidence of sunlight, Nouvel has created an evocative facade, characterized by typical Arab geometries, which transmit to the visitor the magic of the Islamic iconography.

产品设计
PRODUCT DESIGN

i时代建筑设计
DESIGN A BUILDING FOR THE iGENERATION

2050城市设计(再设计)
(RE)DESIGNING TOWNS FOR 2050

1 法则
1 principles

1 法则

产品设计

产品设计是用某个时期现有的材料和技术制作出满足人们需求的产品一个学科。我们日常生活用到的每一件物品实际上都历经了一个设计的过程，这一点我们大多不会注意到。

设计过的产品有以下特点：
- 功用：产品的用途
- 美观：看起来漂亮
- 经济：价格适当

产品可大致归为两大类：
- 简单产品：产品的设计不需要专业技术知识，基本上人人都能做
- 复杂产品：具备专业知识和专业技术的人才能设计出的产品

优秀设计项目的基本要素大致相同，主要有以下几点：

需求
设计的目的是改善生活，满足人们的某项需求。

框架语境
影响产品设计的外部因素主要有：
- 成本：经济资源的情况会影响产品的质量和成本
- 审美：设计师是否跟随时尚潮流的选择也将影响设计产品的审美

时代
产品是设计师所处时代的产物，因为产品是用当时的材料和技术制作出来的，并且与当时的社会环境相容。

1 PRINCIPLES

THE PRODUCT DESIGN

Product design is the science through which we can give shape to man's needs using the materials and technologies available at that particular moment in history. The design process is the conceiving of various objects and consumer goods, which are generally defined products.

The designed objects should be:
- functional: perform the task for which they were designed
- aesthetically appealing: beautiful to the eye
- economically sustainable: have the right price

We can distinguish between two main categories of objects:
- simple objects: those that can be designed without specific skills by almost anyone
- complex objects: those which can only be designed by competent people who have specific knowledge

The basis of a good design project has, as in other fields of design, some basic principles:

NEEDS
The purpose of a design object is to improve our life and to satisfy the needs.

FRAMEWORK CONDITIONS
External factors that affect the design of a product are:
- costs: the economic resources available, which will affect the quality and the final cost of the object
- aesthetics: the need (or not) to follow fashions that will bring the designer to make choices having an impact on the aesthetics in the design phase

TIME
The object of design is the result of the time in which it is conceived; it must be made of the materials and with the manufacturing techniques, which at that moment in history are the most appropriate to achieve the best result in compliance with the framework conditions.

设计过程

无论何种复杂程度，何种类型的产品都必须经过一个富有创意、分段进行的设计过程。

任务和目标

客户需要生产某种产品，并将设计任务交给设计师。客户与设计师共同制定设计目标。

调查研究与市场调研

在这个阶段，设计师和他的团队需要深入市场，了解消费者的实际需求。

调研方式主要有如下几项：

- 与使用者会谈
- 复查已生产相关产品的制作信息和专利情况
- 对产品的相关社会文化进行深入了解，可以从人类学和文化角度进行比较
- 通过调查问卷、小组讨论等，以多种不同的渠道接触到大量产品用户
- 市场分析、零售情况分析、品牌分析这些工作能总结出大量有关产品销售潜力、竞争对手及市场走向的信息

THE DESIGN PROCESS

The creative process that is the basis of good design takes place through a series of stages that are left unchanged for each level of complexity and type of product:

ASSIGNMENT AND OBJECTIVES

The customer expresses his willingness to proceed with the production of an object and gives the assignment to the designer.
Client and designer formulate the objectives and define the framework conditions.

RESEARCH AND MARKET RESEARCH

At this stage, the designer and his team have the task of probing the market to understand what are the actual needs of consumers.

The most common techniques are:

- interviews with users
- review of the literature in order to search for information on products already manufactured and patents
- anthropological and cultural comparisons to analyze in depth the cultural aspects of the society in relation to the object to be designed
- questionnaires, surveys and focus groups: by using various channels at the same time it is possible to involve a large number of users
- market analysis, analysis with regard to retail, research on brand: they provide a great deal of information about the potential sale of the product, the competitors and market trends

分析与评估

收集了大量信息后，规划组将对信息数据进行分类整合，并制定出设计项目的具体细则。
在这个阶段要做到精准地推论信息，并做到以下几点：

- 解释并分析从消费者方得来的信息
- 将消费者的需求和要求分类有序整理
- 按重要等级分类个体的需求
- 明确需求并转换成最后的设计

生成设计项目

经过前两个准备工作阶段，设计师开始正式设计产品。产品的定义是一个创造性的阶段，具有战略性的意义，因为它直接引导设计师构思产品草图和设计理念。

需要开展的活动有如下几项：

- 头脑风暴讨论
- 分析产品特点
- 列出各种创意
- 思维构图

经过以上活动，设计师根据消费者需求和要求确定产品造型。
通过这一过程，囊括了所有的必备信息，产品概念最终形成了。

研发及样品制作

从最初的概念稿开始，设计师必须逐步完善项目细则定义。在此过程中会经过多次的研究和测试。

ANALYSIS AND EVALUATION

After collecting all the necessary information for work, the planning team must organize and catalog all the data in order to reach a definition of the project.
This phase follows a precise reasoning;
it is, in fact, necessary to:

- interpret and analyze the information gathered from consumers
- catalog and organize hierarchically the needs and demands of customers
- establish a hierarchy of importance for individual needs
- define the needs and translate them into the final design

IDENTIFICATION OF THE PROJECT

After the two previous preparatory stages, the designer goes into the actual design of the product.
The definition of the product is the creative phase, a moment of strategic importance as it leads the designer to formulate a draft of the product and the "design concept".

Through activities such as:

- brainstorming
- analysis of the characteristics of the object
- checklist of ideas
- mind mapping

the designer gives shape and body to the needs and demands that have been previously identified.
Through this process the concept, including all the necessary information to develop the project, is born.

DEVELOPMENT AND PROTOTYPING

Starting from the first conceptual draft of the product, the designer must reach gradually a definition of the project.
This process is often characterized by a number of cycles of testing and studies for the completion of the project.

这个阶段必须用到各种工具和技巧，一般来说有如下几种：
- 徒手画稿
- 技术设计
- 利用软件呈现效果图和模拟图
- 模型制作

最后是确定项目中各个部件。

In this stage, it is often necessary to use various tools and techniques, generally:
- freehand drawing
- technical design
- software for the realization of rendering and simulations
- development of models

The end result is the project defined in all its parts, ready to be realized.

验证

VERIFICATION

在这个最后阶段，设计将以三维实物呈现。

In this last phase, the object is finally materialized in three dimensions.

一开始会制作出一件样品，用此来评估设计工作是否正确无误地完成，产品的设计目标是否达成。

At first, a prototype is produced, which has the purpose of assessing whether the design work has been done in the correct way and if the product has achieved the objectives set.

紧接着是第二个分析和研究过程，在此过程中会由第三方对大量的消费者进行调查研究。

Then a second stage of analysis and study follows, which is often characterized by the use of surveys conducted by third parties on a sufficiently large number of consumers.

调研方向如下：
- 产品的安全性
- 是否符合各项标准规定
- 消费者满意度

Analyzed in detail are:
- the object's safety
- regulatory compliance
- the degree of customer satisfaction

如果产品通过了验证阶段，那它将开始大量生产并投放市场。
否则设计师必须回到产品研发阶段中作出必要的项目修改。

If the product passes the verification phase, it is ready to be placed on the market and production will be started. Otherwise, the designer has to return back to the stage of development to make the necessary changes to the project.

生产

PRODUCTION

在生产阶段，即使产品很成功，也有必要再经过一系列的审核修改（如漏洞修正、整体改善、合理化、适应市场要求，等等）。

During the production phase, even if the product is successful, it may be necessary to go through a series of cyclical audits and consequently through changes (bug fixing, improvements in general, rationalization, adaptation to the market, etc.).

职业道德与设计

步入21世纪，人们都已经意识到了人类活动对地球和人类自身将有怎样的影响。

设计师作为一个个体也肩负着社会责任，他工作的根本点应该是为改善世界而做出贡献。

每一项设计过程都应符合职业道德标准，而职业道德标准也是设计框架语境中不可或缺的一部分。

许多不同领域都会参考可持续性标准，产品设计也不例外。

产品设计的可持续性可从三个方面来评估：

- 环境可持续性
- 经济可持续性
- 社会可持续性

环境可持续性

设计工作对资源、废气排放、循环利用、废物处理等方面都会造成一定的影响，因此，设计师作为一名规划者担负着重大的道德责任。

环境可持续性可以从以下几方向展开：

- 利用现成的，可循环利用的材料
- 节约产品生产过程耗费的能源，最好是利用可再生能源
- 产品生命周期中的任何阶段都不能有任何污染排放（生命周期分析（LCA）准确无误）
- 在生产操作阶段保证能源效率

ETHICS AND DESIGN

21 st century man has reached an awareness of the consequences of his actions on the planet and mankind.

The designer, as an individual and a professional with social responsibility, bases his actions on the desire to make a contribution to improving the world.

Every design process must submit to a series of ethical criteria that should be an integral part of the framework conditions of design.

Sustainability issues, to which reference is made every day in many different fields, must be applied to the design of the product.

The sustainability of the product design must be evaluated based on a triple check:

- environmental sustainability
- economic sustainability
- social sustainability

ENVIRONMENTAL SUSTAINABILITY

The designer, as any planner, has a strong ethical responsibility: his work has an impact on resources, emissions, recycling and waste disposal.

Some guidelines related to environmental sustainability are:

- use of readily available materials that are reusable and recyclable
- economical use of energy in the production cycle of the object; the energy used should preferably come from renewable sources
- total absence of pollutant emission into the environment at any stage of product life (accurate assessment of impacts using LCA - Life Cycle Analysis)
- assurance of the product's energy efficiency in the operating phase

经济可持续性

要做到经济可行，生产的产品须符合以下条件：

- 耗能最小化而利润最大化
- 避免依赖有限资源
- 价格合理（满足消费者的需求而获利）

社会可持续性

社会可持续性是指能帮助所有与产品相关的人们改善提高生活质量。

生产过程须确保：
- 对受影响的社区提供帮助和支持
- 工作质量标准细则完善
- 产品生产、使用、废弃过程不会影响相关人们的身心健康

为了让产品环保、经济、社会可持续，在设计阶段就必须考虑以下事项：
- 产品使用的材料
- 产品是否高效
- 产品的耐用性和生命周期

材料

有关材料对环境影响这方面的特点特别需要细心研究，主要有以下几个方面：

- 生物降解材料
- 可循环与已循环材料
- 区分不同的材料

ECONOMIC SUSTAINABILITY

To be economically viable, the product must:

- allow the maximization of profit with a minimum consumption of resources
- avoid relying on limited resources
- have the right price (generate revenues in accordance with the needs of consumers)

SOCIAL SUSTAINABILITY

Social sustainability is the ability to help improve the quality of life of all the people involved in the process.
The production process must ensure:
- support to the affected communities
- adequate standards for the quality of work
- physical and mental health of the people involved in production, use and disposal

To ensure compliance with the environmental, economic and social sustainability of the product, it is necessary to think already during the design phase about:

- the materials to be used
- the efficiency
- the durability and lifetime of the product

MATERIALS

The characteristics of the materials must be evaluated very carefully in relation to their impact on the environment.

In particular, the following must be considered:
- use of biodegradable materials
- use of recyclable and recycled materials
- identification and possibility of separation of the different materials

产品的耐用性和持续性

为减少产品对环境的影响，设计师有必要估算产品的生命周期，即从开发到弃用的时间段。

这个结果可以采用“从摇篮到摇篮”的方式得出。这种方式涵盖了经济、工业、社会各方面的因素。

减少产品带来的危害就要求生产高效的产品。

在保护环境方面，产品的耐用性和持续性是两个最主要的方面。
如果产品可供几代人长时间使用，那么将减少使用新的同类产品而带来的不必要的浪费。

如果产品可以通过维修更新常年使用，资源也可以得到更好地利用。

不容置疑，最“良心的设计”是利用废旧物品的设计。

很不幸的是，这几十年来，人们都倾向于购买低质短命的产品（特别是服装类和饰品类）。

这些压力主要来自于一些大型的销售团体，他们总是以绿色环保为幌子，不计成本地追求利润，最终产生一大堆垃圾，培养了一代不负责任的消费者。

我们相信消费者大众的集体智慧，不久他们也会注重环保意识，并做出改变。

DURABILITY AND LONGEVITY OF PRODUCT

To reduce the environmental impact of the product,
it is necessary to assess the whole life cycle,
from creation to disposal.
This result can be achieved by adopting
the "cradle to cradle" design method.
Cradle to cradle is a holistic process, which covers
the economic, industrial and social aspects of a project.
This attitude results in the development of efficient products
while minimizing the production of waste.

In the environmental perspective, durability and longevity
are two fundamental aspects.
A product that can be used for a long time,
even through multiple generations, minimizes the use
of new objects of the same type, reducing wastage.
If the designer also adds the possibility of renewal and repair
over the years, the result is the abandonment
of "disposable design" and better resource management.

It is undeniable that for most products “ethical design”
is based on the quality of the materials used.

Unfortunately, for a few decades, the buyer has been induced
to orientate towards low quality products (especially
in the clothing and furnishing) with a very short life cycle.

This pressure is created especially by large selling groups,
often hiding behind “ethical-green” campaigns
and environmental certifications while pursuing
monetary gain at any cost, thus creating a sea of waste
and a generation of irresponsible consumers.

We trust that the collective intelligence of consumers
will soon allow awareness of the consequences
of this attitude and a dramatic change in course.

高效产品

评估是否是高效产品的两个主要参数是：

- 能源参数：好的设计首先会考虑降低生产及使用产品带来的能源消耗；设计师在规划阶段就应考虑使用可再生能源。
- 效率：评估产品是否高效的另一方面是看它是否易组装、易使用、易拆卸，是否可以循环利用。

PRODUCT EFFICIENCY

The evaluation of the efficiency of a product relies principally on two parameters:

- energy aspect: the good design project starts by reducing the energy necessary both for the production and for the functioning of the product; the designer should think already in the planning stage about the use of renewable energy sources
- efficiency: the efficiency of a product is evaluated also on its ease of assembly, use, disassembly and the possibility of reuse and recycling.

法则
PRINCIPLES

材料
MATERIALS

加工处理
PROCESSES

2 材料

现如今要开发一个设计项目，有众多种类的材料可供选择。

选择何种材料与以下几个因素相关：

- 产品的用途
- 产品使用所处外部环境
- 设计师想要达到的美观效果

除了选择材料，设计师还必须选择使用何种技术来生产产品。

因此，设计师还必须对所选材料的特性，加工工艺有“360°”（全方面）的了解。

材料的选择必须经过结果分析，并根据设计和所选材料选择正确的生产工艺。

全球生产商们提供了品类众多的材料。

生产产品的主要材料大致有以下几种：

- 陶瓷
- 合成材料
- 人造橡胶
- 玻璃
- 铁制金属
- 非铁金属
- 塑料
- 木材

2 MATERIALS

The designer developing a new project is confronted today with a huge number of possible materials.

The choice of the material to be used is made in relation to:

- the function the product has to perform
- the context in which it is inserted and used
- the aesthetic result that the designer wants to obtain

Parallel to the choice of material, the designer must decide also the technology to be used for production of the object.

It is, therefore, essential that the designer has a “360° knowledge” of the materials, their characteristics and processing techniques.

The materials have to be chosen after consequence analysis, and the correct production technique has to be selected depending on the design and the material used.

The world of production provides a large number of materials.

Some of the most common materials used for producing objects are:

- ceramics
- composite materials
- elastomers
- glass
- ferrous metals
- non-ferrous metals
- plastic
- wood

88 - terracotta

89 - gres

90 - porcelain

陶瓷

陶瓷是取自大自然的泥土，成型后放在窑中烧制而成。

88 陶瓦
89 陶石
90 瓷器

陶瓦
红色或橙色情的黏土，主要用于陶器和建筑用砖、瓦等。

陶石
高温烧制的陶石，一般是灰色和棕色的，主要用来制做陶器、铺砖、建筑板材。这种材制防水且硬度高。

瓷器
高温烧制的有近似玻璃表面的精美瓷器，大部分为白色，主要用于生产昂贵、价值高的产品。

CERAMICS

Materials of natural origin, ceramics are composed of earths that, once molded into the desired shapes, are cooked in an oven.

TERRACOTTA
Red or orange clay, terracotta is mainly used for the production of pottery or construction elements (bricks, tiles, etc.)

GRES
Made by baking clay at high temperature, grès normally has a color between gray and brown and is used for the production of pottery as well as for the paving and plating of buildings, presenting a particular moisture resistance and high hardness.

PORCELAIN
A material with an almost vitrified finishing due to high cooking temperatures, porcelain normally has a white color. It is used mostly for the production of valuable objects.

91 - carbon fiber

92 - fiberglass

93 - multilayer

合成材料

合成材料由一种或多种不同的材料混合制成。
混合工艺的选择主要取决于部件的特性和设计师想要达到的结果。

91 碳纤维
92 纤维玻璃
93 复合材料

碳纤维

碳纤维是由片状石墨微晶等碳纤维沿纤维轴向束制并混合热固树脂而成的。碳纤维材料重量轻但强度高，因此它主要用于一些高强度产品。

纤维玻璃

纤维玻璃由热固树脂和玻璃纤维合成，抗腐蚀性能极强。纤维玻璃广泛应用于航空、船舶等经常与腐蚀性介质接触的领域。

复合材料

这种材料是用粘胶的方式将多层材料粘合在一起。
多层材料可以是木材复合，也可以是航空领域的合成材料，它们都具有重量轻，强度高的特点。

COMPOSITES

Composite materials are made by the coupling of one or more materials that are different from each other.
The coupling technology is chosen according to the desired result and the characteristics of the individual components.

CARBON FIBER

Carbon fiber is a material produced by the coupling of braided leaves of thin carbon fibers with a matrix of thermosetting resin.
Carbon fiber has an extraordinary high traction resistance to weight ratio; for this reason, it is used for high performance products.

FIBERGLASS

Fiberglass is produced by the coupling of thermosetting resins with glass fibers; the material has excellent corrosion resistance.
Fiberglass is used in the worlds of aeronautics, shipbuilding and in all settings where products stay in contact with corrosive agents.

MULTILAYER

This group of materials consists of all those realized by the coupling of layers of material by means of adhesive films.
It ranges from simple wooden plywood up to the composite materials used in the world of aeronautics, with a light weight and high performance.

弹性材料

弹性材料是天然高分子聚合物或人造橡胶，弹性好。

94 天然橡胶
95 硅胶

ELASTOMERS

The elastomers are polymers of natural or artificial origin, characterized by elastic properties.

94 - natural rubber

95 - silicones

天然橡胶

从橡胶树树液中提取的弹性材料。
橡胶经过处理加工可以制成各种物品，而且弹性极佳，如橡胶手套、橡胶章印。

硅胶

硅胶是在人工材料基础上制成的高分子材料。硅胶系列材料可以有多种形态（既可是油状，也可以是橡胶那样的比较稳定的形态），这种材料耐高温，耐磨损。

NATURAL RUBBER

A flexible material that is derived from the sap extracted from the rubber tree.
Once processed, it is used to produce objects of any kind characterized by a high elasticity, such as gloves and elastic seals.

SILICONES

Polymer based materials of artificial origin.
Silicones are a family of materials with varied consistency (from oily to rubbery); they are particularly resistant to high temperatures and wear.

玻璃

玻璃是经过含硅砂熔烧制成的。
玻璃熔液中添加不同的化合物能让玻璃拥有不同的特点和颜色。
玻璃透明度好，造价成本相对较低，而且可以回收利用，因此广泛用于各个领域。

96 玻璃制品
97 玻璃片

GLASS

Glass is obtained by the fusion of sandy materials based on silica.
Glasses with the most different characteristics and colors can be obtained by the addition of other compounds.
Thanks to its characteristics of resistance and transparency, its competitive cost with other materials and the ease of recycling, it is widely used in the most disparate sectors whenever transparency is needed.

96 - glass objects

97 - sheet glass

98 - cast iron

99 - carbon steel

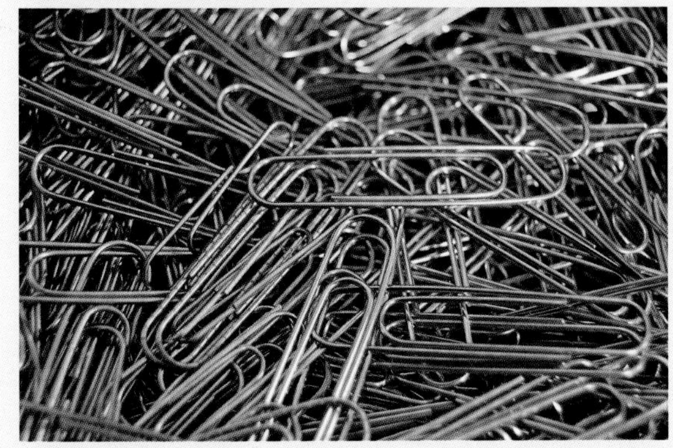
100 - stainless steel

铁制金属

铁合金是以铁为主要元素的材料，强度高，应用普遍。
铁合金中碳的含量影响材料的结构特性。

98 铸铁
99 碳素钢
100 不锈钢

铸铁
铸铁是一种又硬又脆的金属，很容易生锈，广泛应用于工业领域。

碳素钢
碳和铁的合金，受热后有弹性，硬度强度高，特别适合制作工具。

不锈钢
不锈钢具有很强的耐腐蚀性，多用于露天产品，因其外观光亮，这种材料也备受青睐。根据合金成份构成比例，不锈钢也会有不同的特点。

FERROUS METALS

The metal alloys based on iron, appreciated for their strength, have become fundamental within our society.
Their mechanical properties vary according to the percentage of carbon present in the alloy.

CAST IRON
Hard and brittle metal that rusts easily, it is suitable to a wide range of industrial applications.

CARBON STEEL
An alloy composed of iron and carbon, it has a high hardness and mechanical strength that makes it particularly suitable for the production of tools.

STAINLESS STEEL
Stainless steel is characterized by a high resistance to corrosion; it is used when the product stays in contact with atmospheric agents, and is also appreciated for its aesthetic brilliant finish.
The characteristics and properties of stainless steel vary according to the alloy composition.

101 - copper

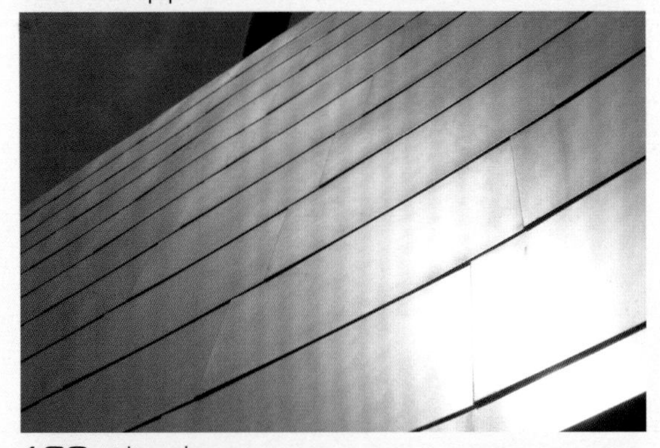
102 - titanium

103 - gold

非铁制金属

非铁制金属是不含铁、而用其他金属合成的材料。

非铁制金属通常是从不同的矿石中提取的，加工处理复杂耗时。

非铁制金属成本非常高，因为加工这种材料需要大量能源，矿石也很稀有。

每种非铁制金属有着各自特点，主要从以下几个方面来概括：

- 亮度
- 抗氧化和抗腐蚀能力
- 导电导热能力
- 硬度
- 熔点
- 柔韧性

101 铜
102 钛
103 金

NON-FERROUS MATERIALS

Non-ferrous metals are metallic materials that do not contain iron, but are made from other metals or from alloys thereof.
In general, the non-ferrous metals are extracted from various minerals and are processed by means of long and complex processes.
The cost of most non-ferrous metal materials is high, both because of the burden of the production processes that often involve the use of large amounts of energy and for other causes related to the supply of the mineral.

Each non-ferrous metal is characterized by specific features, the most significant of which are:

- lightness
- resistance to oxidation and corrosion
- thermal and electrical conductivity
- hardness
- fusion point
- malleability

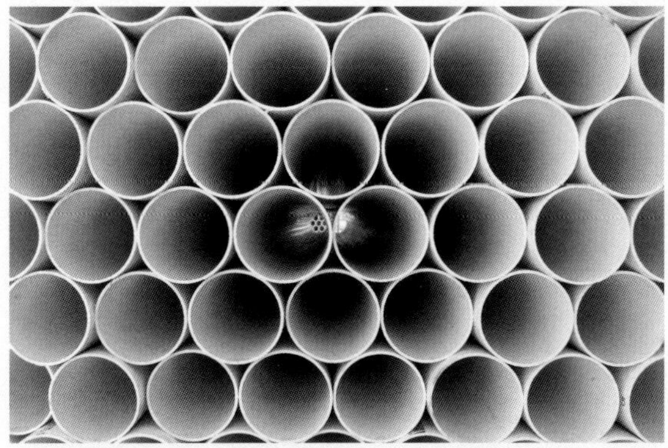
104 - polyvinylchloride (PVC)

105 - polyester (PET)

106 - polystyrene (PS)

塑料

塑料是从石油合成物中加工而成的高分子材料。首次发现塑料是在1862年，之后塑料越来越广泛地应用于我们日常生活中。

塑料易生产，易加工，成本低，因此现在广泛用于各个领域。

塑料大致可分为两大类：

- 热塑性聚合物 —— 易变软融化 —— 如聚碳酸酯(PC)、聚乙烯(PE)、聚丙烯(PP)、聚氯乙烯(PVC)
- 热固高分子聚合物 —— 受热变硬 —— 如环氧树脂、聚胺酯 (PU)

近几年，从植物中提取的浆粉开始走入市场，不久的将来还会有更大发展。

104 聚氯乙烯 (PVC)
105 涤纶，聚酯 (PET)
106 苯乙烯聚合物 (PS)

PLASTICS

Plastics are synthetic polymers that are derived from oil.
Plastic was discovered for the first time in 1862 and since then has had an ever-increasing role in our lives.
Today, plastics are used in various fields of application; they are particularly appreciated for their ease of production, processing and low cost.
Plastic materials are generally readily recyclable and, even if they are not, may be used for other purposes o r burned to produce energy.
Plastics can be divided in two big macro families:

- thermoplastic polymers - which soften and melt - such as polycarbonate (PC), polyethylene (PE), polypropylene (PP), polyvinyl chloride (PVC)
- thermosetting polymers - that harden with heat - for example, epoxy resins and polyurethane (PU)

In recent years products made with bio-based plastics, derived from vegetable starches, are entering the market, and are believed to have a large resonance in the near future.

107 - wood

108 - wood objects

109 - wood objects

木材

木材是人类最早使用的材料之一。
树木的种类繁多，木质结构和外观质地各不相同，使用的用途也不尽相同。
在日常生活中，不经意间我们就会在家中或在其他场所使用到木材。
木材绝对是天然可循环的材料。

107 木材
108 木材制品
109 木材制品

WOOD

Wood is one of the first materials that man learned to use.
It comes in a huge variety of d ifferent wood species, characterized by different mechanical and aesthetic qualities that make each type of wood suitable for particular uses.
In everyday life, even unintentionally, we make widespread use of wood within our homes and in real estate in general.
Wood is completely natural and recyclable.

法则
PRINCIPLES

材料
MATERIALS

加工处理
PROCESSES

3 加工处理
3 processes

3 加工处理

如今的加工工艺种类繁多，日新月异。

选择何种加工工艺来完成项目取决于所选的材料。

为达到最理想的效果，每种材料都必须用最合适的制作加工方式来处理。

一般来说，加工处理有五项基本的简单操作：
- 去除
- 添加
- 压印
- 塑型
- 延展

每一项操作包含无数个精细操作。另外，产品完成后还有外表涂层工艺。

产品加工过程可分为以下几个功能类别：
- 切割
- 连接
- 融合
- 制模
- 抛光

3 PROCESSES

The processing techniques available today are varied and quickly evolving.
The right processing technique to be used for the realization of a design project goes along with the choice of material.
To get the best result, each material must be processed with a suitable manufacturing practice.

There are basically five ways to achieve an object, which correspond to five simple basic actions:
- remove
- add
- print
- fashion
- expand

Each category comprises countless variations.
There are also processes that concern only the surface finishes and the product's coating.

The processes for obtaining products can be grouped in the following functional categories:
- cut
- joint
- merge
- mold
- finish

110 - laser cutting

111 - milling

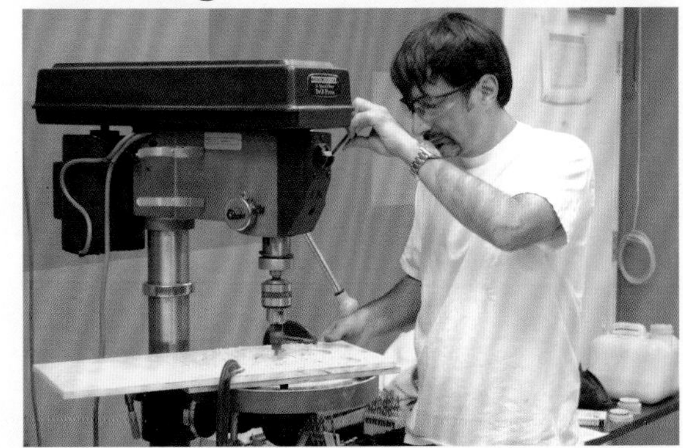
112 - drilling

切割

切割是用外力将材料分成不同的部分。

切割方式有如下几种类型：

- 割槽：在片材上进行割槽
- 锯：用转动或滑动的刀片切割
- 冲切：用做成一定形状的刀片来切割一些较薄的东西，如纸片、纸板、布片
- 水切割：利用高压水来切割，这种切割不会产生热量，所以也应用于多种材料
- 激光切割：用非常细的激光束来切割金属和非金属，但是会产生热量

一些机械加工技术如：
- 钻
- 铰
- 铣
- 修形
- 车削

在广义的认识上这些技术也属于切割

110 激光切割
111 铣
112 钻

CUT

The cutting operation consists of the separation of a material into more distinct parts by the application of a force.

There are various types of cutting:
- trenching: performed in a mechanical way on materials in sheet
- sawing: performed with the use of rotating or sliding blades
- die cutting: uses a die (shaped cutting edge) to cut through low strength materials such as paper, cardboard and cloth
- water jet cutting: cuts using high-pressure water, it is used on a variety of materials and does not produce heat
- laser cutting: cuts metallic and non-metallic materials using a thin laser beam; it produces heat

A number of machine processing techniques such as:
- drilling
- reaming
- milling
- shaping
- turning

also belong to the category of cutting in the broad sense.

113 - welding

114 - adhesive bonding

115 - mechanical junction

连接

连接类型大致包括机械连接、结构连接、化学连接。通过连接小的部件可以组成更大的物品。

不同的材料和技术需要用到不同的连接，如：

- 焊接：以加热高温或者高压的方式接合金属，必要时可加熔填物辅助
- 粘连：通过黏合剂（胶水或胶带）将两种元素接合在一起。这种方式可用于多种不同的材料
- 机械连接：利用铁钉、螺丝等连接不同的部分
- 硬焊：利用熔化的焊接接头将金属接合在一起

113 焊接
114 胶粘
115 机械连接

JUNCTION

The junction consists of the mechanical, structural or chemical union of several smaller parts, with the purpose of forming larger size objects.

There are different types of joints that exploit different materials and technologies, such as:

- welding: joining metal parts by means of heat or pressure, with or without filling material
- adhesive bonding: it involves gluing two elements using an adhesive (glue or tape); the method of adhesive junction can be used on many different materials
- mechanical junction: assembly through the use of a wide range of fastening systems such as nails, screws, etc
- brazing: technology used to bond metals, consisting of the junction of two components using a metal filler that is melted

融合

融合是将材料熔解成液态再注入模具中。
待材料冷却后便成为模具的形状。
这种融合工艺适用于多种材料（金属、玻璃、塑料，等等）。

铸造加工有以下几种类型：

- 注模：将颗粒状高分子材料加热或加压注入模具中
- 吹塑法：玻璃工业中制造瓶形产品常用的加工方法
- 旋转铸塑：适合铸塑陶瓷类空心容器
- 压铸法：熔化的金属材料被压入模具中
- 铸造成型：与压铸法相似，主要用于陶瓷材料
- 蜡模铸造：用蜡制的模型来塑型

金属材料特有的锻造和旋转操作也可归入融合工艺类。

116 浇铸
117 玻璃吹塑

FUSION

Fusion consists of pouring a liquid material inside a mold.
When cooled, the material takes the shape of the mold in which it was placed.
Merger techniques are used for all materials (metals, glass, plastics, etc.) that merge with an increase in temperature.
There are various types of casting processes such as:

- injection molding: grains of polymers are subjected to heat and high pressure and then injected into the mold
- blow molding: typical process of the glass industry using air or gas for the formation of bottles
- rotational molding: ideal for the formation of hollow shapes from ceramic type materials
- die casting: the molten metal material is pushed inside the molds using pressure
- forming by casting: it is similar to die casting but it is used mostly for materials of ceramic type
- lost wax casting: casting in a mold made from a wax model

Also classified as "fusions" are forging and spinning operations that are specific for the processing of metals.

116 - forming by casting

117 - blowing glass

118 - keying

119 - thermoforming

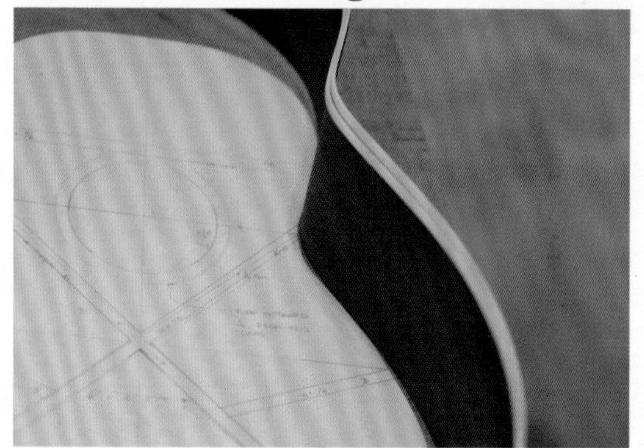
120 - steam bending

成型

成型加工可以制作出金属片、管道，以及市场常用的标准尺寸的形状。

主要的成型工艺有如下几种：

- 折叠法：使用折叠机将二维平面的元素转化成三维立体的形状
- 键控法：专业工匠通过机械力将材料打造成任意形状
- 模具成型法：用两排钢压制中间的金属片来塑造形状
- 热力塑型：主要用于塑料，受热后通过真空压力注入模型
- 蒸气弯曲：特别适用于木材片的弯曲成型。

118 键控法
119 热力塑型
120 蒸气弯曲

FORMING

The forming processes include a series of operations that allow the shaping of materials from metal sheets, pipes and pieces of basic shapes and sizes easily available on the market.

The main techniques are:

- folding: through the use of folding machines, two-dimensional elements are transformed into three-dimensional shapes
- keying: specialist artisan process that allows us to obtain any form by the application of mechanical force to the material
- die forming: process used to shape metal sheets to obtain any form by pressing them between two arrays of steel
- thermoforming: technique applied to plastics that are heated and adhered to a mold by vacuum depression
- steam bending: process particularly suitable for bending wood laminates.

抛光

在加工的各个环节中要预先考虑抛光层的处理。

抛光既可以让产品美观也可以防止材料被腐蚀。

需要抛光的层面必须经过研磨喷洗处理干净，这样才能接融最后的涂层。

抛光操作主要有以下几项：

- 电镀：主要是在物品表面覆盖金属层（常用的是铬）
- 喷涂：喷涂漆料
- 粉末涂层：将粉末涂料涂在物品表层，再放入炉中烧制

121 镀铬
122 涂料喷枪

FINISHES

The finishing layers that can be applied to materials have to be considered in the range of possible processes.
The finishing layers may be applied both for aesthetic reasons and to prevent the effects of corrosion on the material.
Before carrying out finishing processes, it is important to ensure that the areas to be treated are cleaned with abrasives or blasting in order to receive the final substrate correctly.

The main operations of finishing are:

- plating: it consists of covering a surface with a metal (typical example is chrome)
- spray coating: applying a layer of paint
- powder coating: process used for the coating of materials through the application of a powder, which is subsequently cooked in an oven

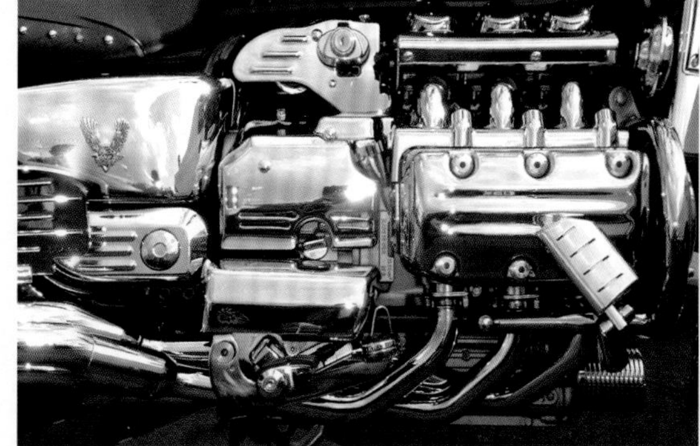

121 - chromium plating

122 - spray painting gun

1 设计法则
多学科方法
三大支柱要素
如机器般精密的建筑

1 principles
multidisciplinary approach
the three pillars
the building as a machine

1 设计法则

艺术和建筑设计是不同的范畴

建筑设计本身没有一个完全纯粹的界定，它总是跟某种功用相关联。

现在在建的很多建筑都不能认为是建筑设计，而是纯艺术：
这些建筑具有标志性，但与建筑设计却毫不相干。
建筑设计工作是设计能满足人类居住需求及其他功能需求，可供人类使用的建筑，涉及审美学，成本。

建筑外墙与功能之间的呼应

现在的建筑通常首先考虑的是外在审美，而不是考虑它的可用性和内部空间功能。
这样一来，建筑的使用者不得不住进这个有所缺失的空间，并做出负面的反馈。
在设计和建造的过程当中，建筑师应该考虑建筑的外形和内部功能的匹配，反之，外部造型须反映出建筑的内部功能。

1 PRINCIPLES

ART AND ARCHITECTURE ARE SEPARATE ENTITIES

Architecture never has a meaning by itself alone,
it is always linked to a function.

Many of the buildings that are being built today
cannot be considered architecture, but pure art:
they accomplish a symbolic function but
have nothing to do with architecture.
An architectural work serves the function of shelter
for humans, responds to requirements of functionality,
aesthetics and cost and must be usable by humans.

CORRESPONDENCE BETWEEN THE SHELL AND THE FUNCTION

Buildings nowadays are often conceived while primarily thinking
about the aesthetic of the outside, without considering
the usability and the functions of the interior space.
This is reflected in a negative way to the user, who
as a consequence has to live in inadequate environments.
During the creative and constructive process,
the architect must be able to match the appearance
of the outer shell with its function and vice versa;
the shell must reflect the building's function.

建筑
ARCHITECTURE

艺术
ART

多学科方法

利用多学科方法是成功设计的基础。

现今如果没有其他学科的专家合作，要完成一项建筑设计几乎是不可能的。
在一个团队里，建筑师负责主导团队及协调各科知识。

建筑师不能做其他专家的工作，但他必须对设计相关领域的知识有所了解，而且有能力领导和协调团队，探索出完成目标的最佳方案。

多学科方法的独特结合必定能有助于设计的圆满成功。

不了解各学科的基础知识、基本原理和技术法则，就无法规划、构想、完成一项设计。

MULTIDISCIPLINARY APPROACH

Fundamental to good design is the multidisciplinary approach.
Nowadays it is impossible to think of realizing an architecture project without the cooperation of other specialized professionals.
Within a team, the architect plays the role of director and coordinator of multiple knowledges.
The architect must not do the work of the other specialized technicians himself, but having the base knowledge relating to all areas that affect the design and being able to lead the team, must be able to coordinate and exploit the various professionals in the best way to achieve the common goal.
The multidisciplinary approach combined with creativity contribute to the success of the project.

Whoever does not know the basics, the principles and the technologies, is not able to plan, to conceive and implement.

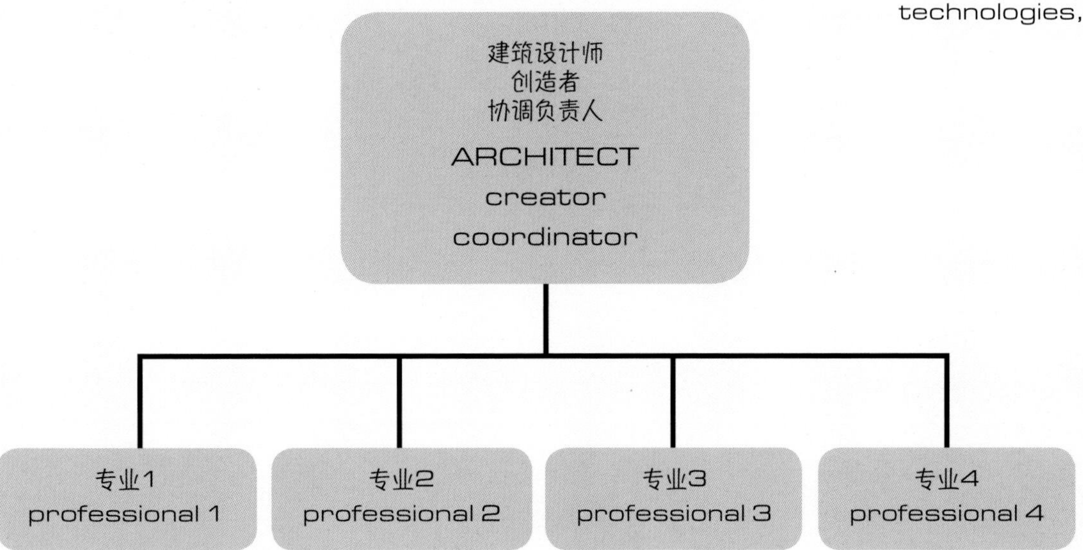

三大支柱要素

地点

建筑物都是座落在某一个地点。所以建筑设计必须考虑该地点的地形、气候、环境因素等特点。

时间

建筑物必须是“时代的产物”。
从它的审美特点、运用技术方面都能反映当时的时代。

用户

建筑物不是一个容器，它是为了让人居住其中。所以设计必须考虑居住者的需求、兴趣和习惯。

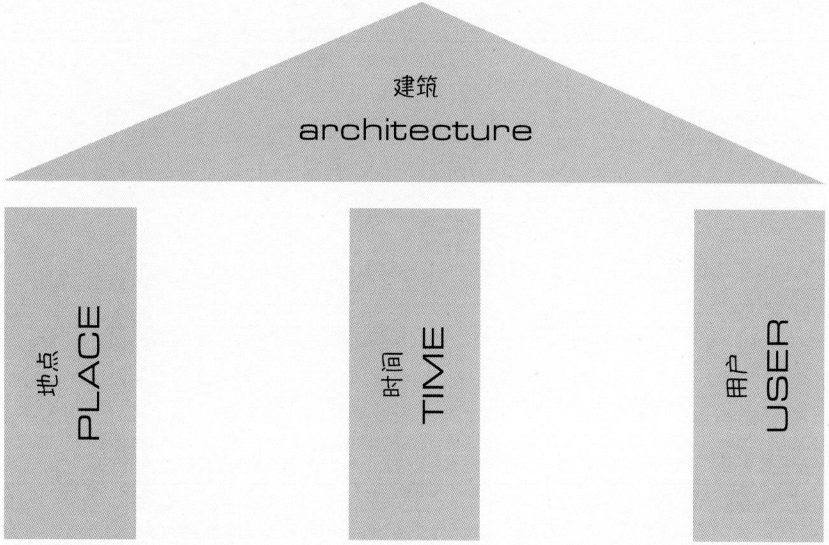

THE THREE PILLARS

PLACE

Architecture must arise from the place where it stands.
The project must be adapted to the characteristics of the site, to its morphology, to its climate and to other environmental variables.

TIME

A product of architecture must be a “child of its time”.
The project must be an expression of the moment in which it is conceived, regarding both aesthetics and the up to date technologies.

USER

Architecture is not just an empty container:
it is made essentially by the people who live in it.
The project must be able to respond to the needs, interests and habits of those who will enjoy it.

建筑
architecture

地点
PLACE

时间
TIME

用户
USER

协调一致的原则

在设计初期建筑师就必须以三大支柱要素为出发点。
设计师必须在地点、时间、用户需求方面有全盘的考虑。
只有充分考虑到这三个要素，设计才能与当地的景观、环境、历史时代协调一致，
给用户带来和谐协调感。

PRINCIPLE OF COHERENCE

The task of the architect is to respect the three pillars
from the early design phase.
The designer must be consistent with thc place,
time and user's needs.
Only when based on these three fundamental pillars
will the project end up being respectful of the landscape
and the environment, in harmony with man and his time.

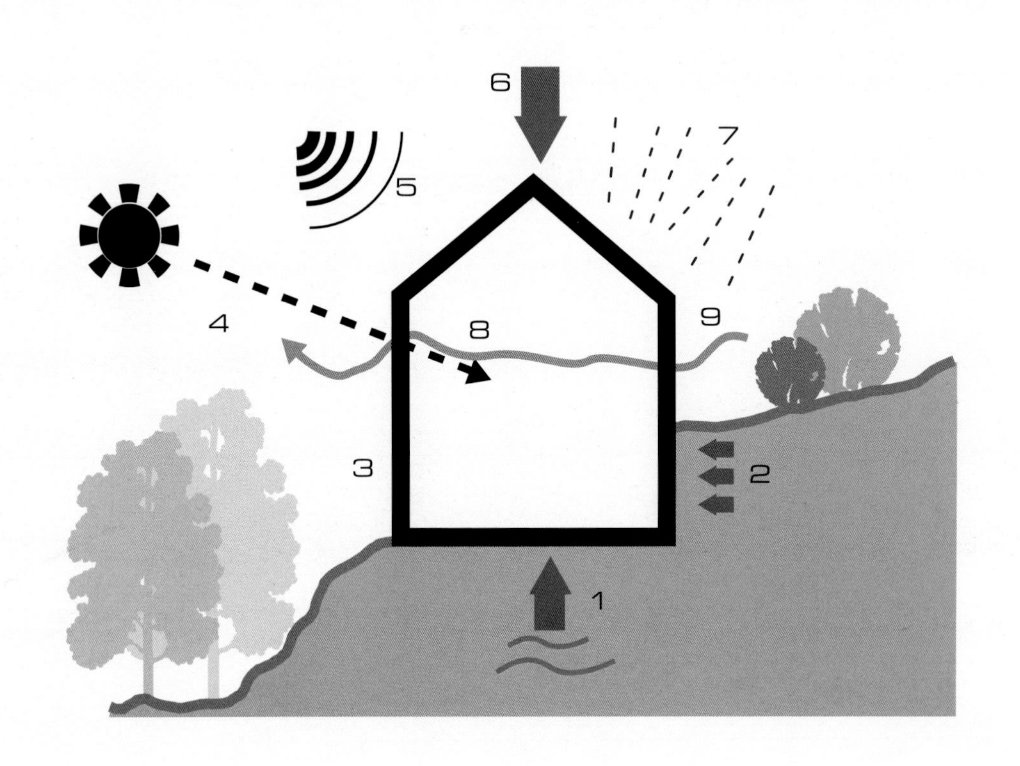
6
7
5
4
8
9
3
2
1

COMFORT

如机器般精密的建筑

建筑物可以被当成一架复杂的机器，这样的机器融合了各种不同的系统、特点和技术，目的就是为了满足人们的两项基本需求：对外遮风蔽雨；对内舒适方便。

设计师的工作就是评估各项变量因素，选择最适合的技术和系统。

建筑物的外表及其特点

- 隔离湿气和氡 (1)
- 抗地表压力 (2)
- 出入口管理和安全 (3)
- 控制日照和阴影 (4)
- 御寒
- 降噪 (5)
- 抗压 (6)
- 避风雨 (7)
- 保障自然通风 (8)
- 控制采光 (9)

内部舒适便利的需求

- 调节光线量
- 电力布线
- 水力布线
- 收集、发布、处理来自各个系统的信息

THE BUILDING AS A MACHINE

The building can be considered as a complex machine that integrates different systems, features and technologies to allow people to answer two fundamental needs: shelter from the external environment and guarantee of internal comfort for the user. The designer's task is to evaluate the various factors, choosing the most appropriate technology and systems.

OUTSIDE, FEATURES OF THE BUILDING'S SKIN

- prevent the entry of humidity and radon (1)
- withstand the pressure of the soil (2)
- manage access and security (3)
- control solar radiation and shading (4)
- protect from cold
- exclude noise (5)
- withstand the loads (6)
- protect from the elements (7)
- ensure natural ventilation (8)
- manage the light (9)

INSIDE, REQUIREMENTS FOR COMFORT

- adjust the light
- distribute electricity
- distribute water
- collect, distribute, manage information from the different systems

2 结构工程
垂直方向承载结构
基础
梁和板

2 structural engineering
vertical load-bearing structures
foundations
lintels and slabs

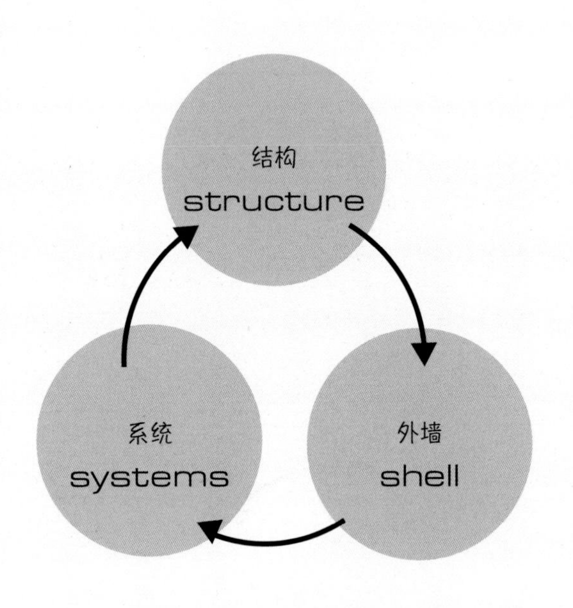

2 结构工程

结构工程是一门研究工程结构受力和荷载的学科。

建筑设计过程是一个选择合适材料和建筑方法的复杂决策过程。

设计师必须在以下三个领域做出互不冲突的选择：

- 结构：处理受力和荷载
- 外墙：建筑物和环境之间的关系
- 技术系统

我们分三个方面来讲结构工程：

- 垂直方向承载结构
- 基础
- 梁和板

2 STRUCTURAL ENGINEERING

Structural engineering: science of the management of forces and loads.

The architectural design process unfolds in a complex decision-making process through the choice of appropriate materials and building methods.
The designer must be able to make consistent choices in three areas:

- STRUCTURES: management of forces and loads
- SHELL: relationship between building and environment
- TECHNICAL SYSTEMS

In the context of structural engineering, we can split it up into:

- vertical load-bearing structures
- foundations
- lintels and slabs

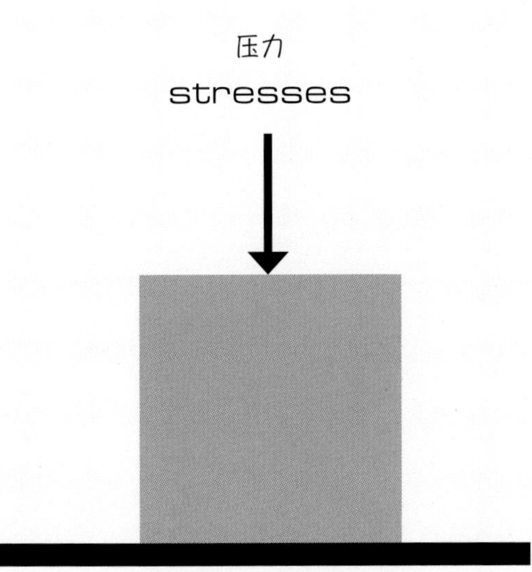

垂直方向承载结构

垂直方向承载结构能把建筑物产生的压力传递到地面：

- 永久荷载
- 活荷载
- 积雪
- 地面沉降
- 动态作用

垂直方向承载结构必须能支撑整个建筑，保证居住者的安全，保持平衡和耐力，抗变形。

VERTICAL LOAD-BEARING STRUCTURES

Vertical load-bearing structures transmit the stresses produced by the forces acting on the building to the soil:

- permanent loads
- operating loads
- snow
- thrust of the ground
- dynamic effects

Vertical load-bearing structures have to support the building and ensure the safety of occupants in every moment and condition by maintaining balance, endurance and deformability.

垂直方向承载结构可以分为以下三种类型：

- 核心筒式
- 墙体式
- 柱式

The vertical load-bearing structures can be divided into three main typologies:

- cells
- septums
- pillars

	类型 - type	重量 - weight	灵活性 - flexibility	承重 - load	稳定性 - stability
A	核心筒式 cells	高 high	低 low	均布荷载 uniform	稳定 stable
B	墙体式 septums	中 medium	中 medium	水平荷载 oriented	大体稳定，某一方向稳定 on average stable, to be stabilized in a direction
C	柱式 pillars	低 low	高 hight	集中荷载 punctual	不稳定，须有两个方向稳定 unstable, to be stabilized in two direc-tions

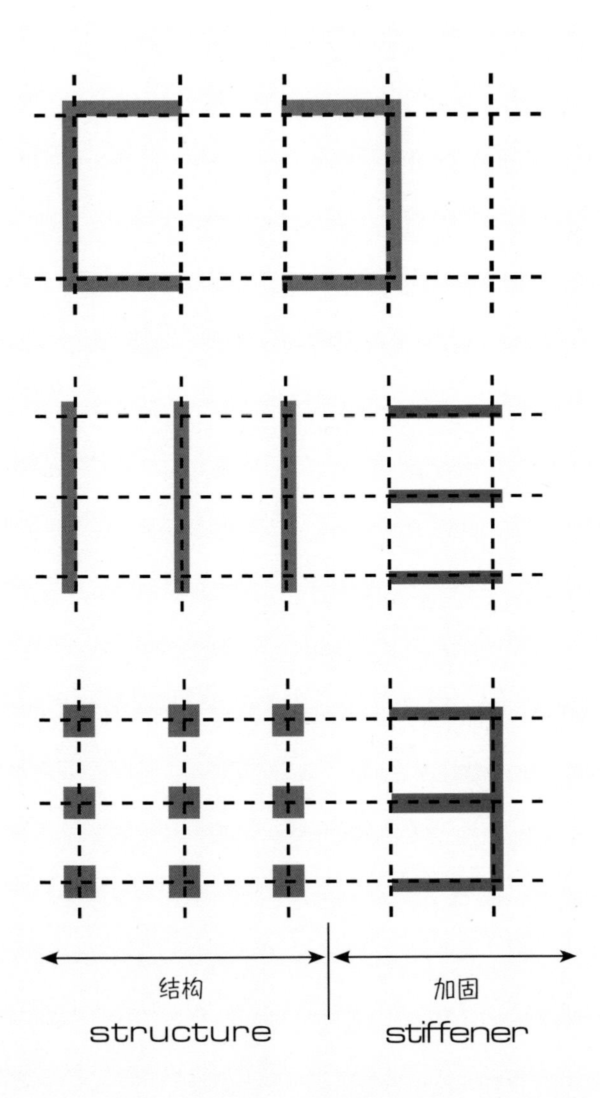

以上三种结构各有利弊。

核心筒式和墙体式比柱式更稳定，但更受局限。

柱式结构更灵活，但是比起前两者来却不太稳定。

因此，我们可以采取柱式、墙体式和核心筒式组合的结构。

如果采用柱式作为垂直方向承载的基本结构，那就必须要有一个二级支撑体系（应力刚化单元）来吸收水平方向的压力。

The three different types of structures have each both positive and negative aspects.
Cells and septums are more stable than the structures with pillars, but more binding.
The pillar structures are more flexible but also more unstable than the previous two.

Hence the need to resort often to a mixed solution: adopt pillar structures (hence flexible) and stabilize using septums or cells, which have to be properly positioned.

The use of pillars as the basic structure (for vertical loads) inevitably leads to the adoption of a secondary supporting structure (stiffening element) that has the purpose of absorbing horizontal stresses.

地面荷载方式

LOADS CONVEYANCE TO THE GROUND

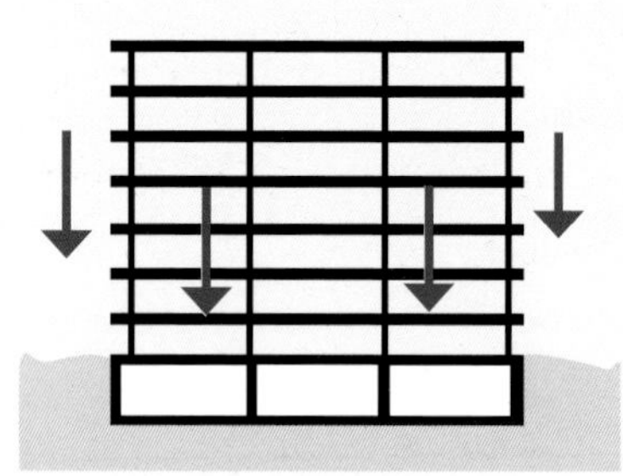

1.

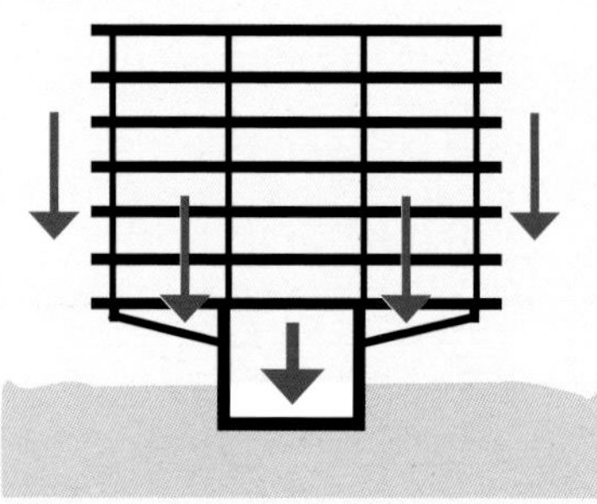

2.

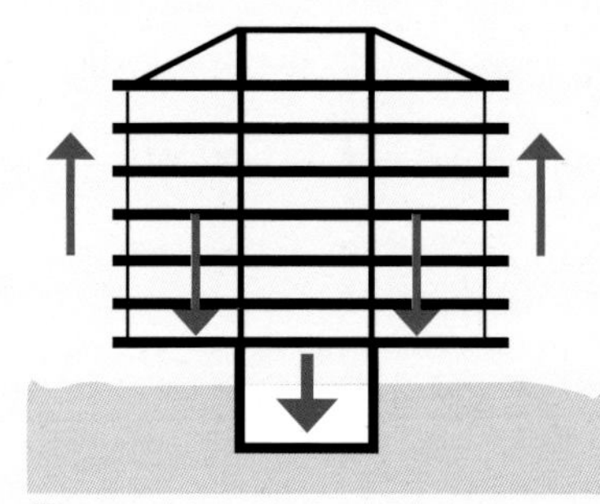

3.

基础

垂直方向的荷载通过基础被传递到地面。

为了防止结构受力不均而导致建筑物受损，基础对地下的压力应分布均衡。

基础的荷载主要来自于建筑物的自重，以及其他因素带来的荷载，如用户、偶发荷载、积雪，等等）。

选择基础体系的决定因素是：

- 地面
- 荷载

1. 通过支撑柱将荷载传递到地面
2. 通过地下的结构柱传递荷载
3. 通过顶层张力使侧向荷载返向，荷载集中到建筑物中心部分并被传递到地面

FOUNDATIONS

The load taken by the vertical structures is transferred to the ground through the foundations. To prevent irregular settlement of the structures - and thus damage to the building - foundations must exert a uniform pressure on the ground below.

The load that is applied to the foundations, is given by the sums of the weight of the building plus all the loads acquired (users, accidental loads, snow, etc).

The factors determining the choice of the system of foundations are:

- the ground
- the applied load

1. loads conveyed to the ground by compressed pillars
2. loads conveyed on a structure below: compressed pillars
3. resumption of lateral loads by elements under tension to the top floor; the loads are concentrated in the central part of the building and transferred to the ground by compression of the structures

地面

按土质层、沉积层不同的材料，地面被分成几种不同的类型。

地面的承重能力可根据其结构和自然地质来测定。

地面类型主要有以下几类：

- 低抗力土层（黏土）
  10 ton/m²
- 中度抗力土层（沙质）
  25 ton/m²
- 抗力土层（沙砾）
  50 ton/m²
- 大理石
  150 ton/m²
- 花岗岩
  300 ton/m²

在规化的初始阶段就必须通过地质勘查对地面情况进行研究。
通过分析我们可以了解到：

- 土层的类型
- 地下水位表
- 附近的水纹活动
- 地震带类型
- 山形地带的山崩和泥石流情况
- 地表下陷的可能性

THE GROUND

There are several different types of grounds, formed by layers, sediments, different materials and quarrels.
The bearing capacity of the ground depends on their structure and geological nature.

The typologies of ground can be classified mainly in:

- soils with low resistance (clay)
  10 ton/m²
- soils of medium strength (sand)
  25 ton/m²
- resistant soils (gravel)
  50 ton/m²
- marble
  150 ton/m²
- granite
  300 ton/m²

In the initial phase of planning it is necessary to proceed with a careful study of the ground by geological surveys and studies.
From the analysis, we can see:

- the type of soil
- the level of the underground water table
- the behavior of nearby waterways
- the type of seismic zone
- the danger of landslides and avalanches (mountainous terrain)
- the possible risk of ground subsidence

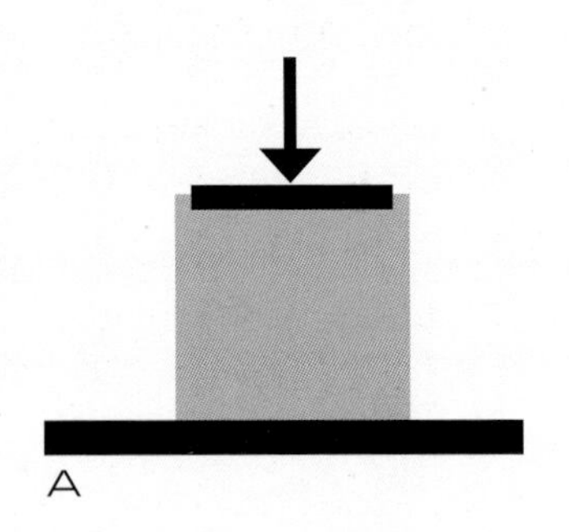

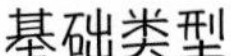

基础类型

基础类型大体可以分为两类：
A 浅基础
B 深基础

A - 浅基础

基础表面由土层的承重能力来定义。

浅基础还可细分为：

- **平台基础**
  这种类型的基础由基础面和平台组成。
  平台基础通常用来承受高度重力，或者是低抗力土层，或是低楼层的荷载需求。

- **条形基础**
  这种基础通过一个连续的线性面传递荷载。

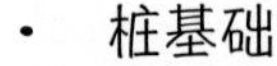

- **桩基础**
  主要用于浅抗力土层和框架结构的建筑。桩的形状根据它的功用来决定。荷载的传递是一个渐进渐宽的过程，最后将荷载传递到较大面积的地表。

TYPES OF FOUNDATION

Foundations can be divided into two macro families:
A shallow foundations
B deep foundations

A - SHALLOW FOUNDATIONS

The foundation surface is defined by the bearing capacity of the soil.

Shallow foundations can be classified into:

- **PLATE FOUNDATIONS**
  They constitute both the foundation and the decking.
  Plate foundations are used for high loads or on soils with low resistance, or if there is the need for a load-bearing floor in the lower floor.

- **LINEAR FOUNDATIONS**
  They transmit the load by distributing it on continuous linear surfaces.

- **FOUNDATIONS ON PLINTHS**
  They are used on resistant soils to shallow depth and in buildings with a frame structure.
  The shape of the plinth is determined by its function, it must transmit loads by gradually widening the section of the pillar in order to transmit the load on a large enough area of ground.

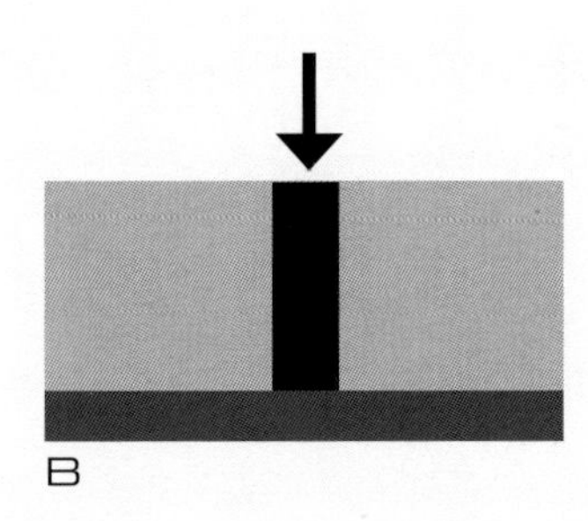

B - 深基础

基础的深度取决于土层的连续性和承重土层的厚度。如果土层抗力低于建筑物传递的荷载，或是抗力土层所处位置较深，那么就有必要采用深基础。

深基础可分为以下几种：

- 灌柱桩
  灌柱桩将荷载传递到抗力土层以下，它们在特定的地点灌铸成型，起到的作用就像立于地表抗力土层的桩柱一样。

- 打桩
  与灌柱桩的工作原理相同，只是桩是靠外力打压进土层的。

- 悬浮桩
  如果抗力土层太深，无法达到，那么只能通过增加承重面积，多桩集中在一起来稳固地表。

B - DEEP FOUNDATIONS

The depth of the foundation is defined by the soil consistency and the depth of thc bcaring layers.
Deep foundations are necessary when the soil has a low resistance compared to the load transmitted from the construction or when the resistant soil is located at a considerable depth.

Deep foundations can be classified into:

- CAST PILES
  They are used to transmit the load to the underlying layers of resistant soil.
  They are cast in place and function like real pillars resting on the resistant layers of the ground.

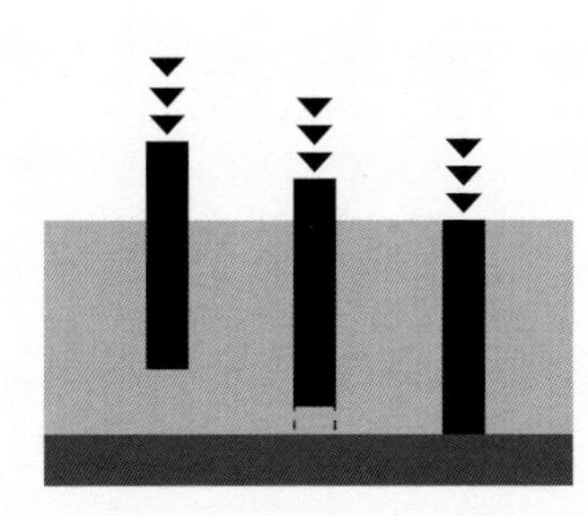

- BEATEN PILES
  They work on the same principle of cast piles, but are placed in the ground by beating.

- SUSPENDED PILES
  They are used when the resistant soil is located too deep to be reached.
  This solution aims to consolidate the soil in the ground by placing the piles close together, thus increasing the load-bearing capacity.

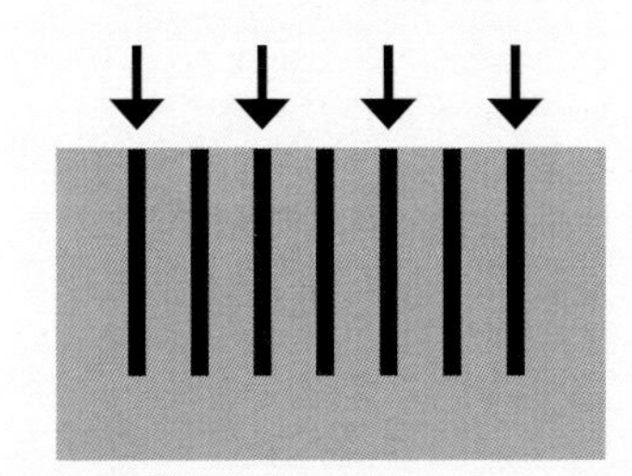

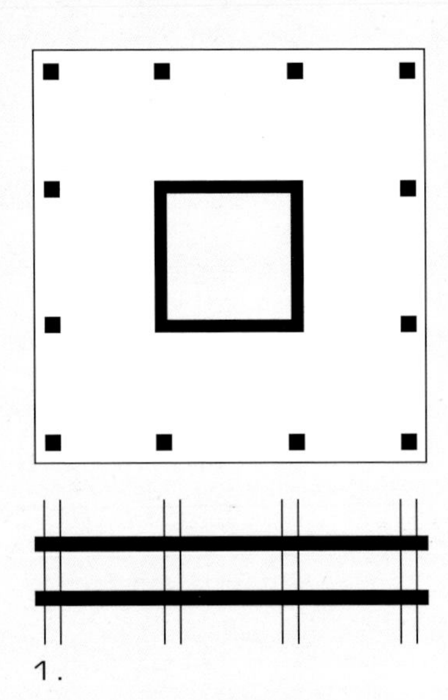

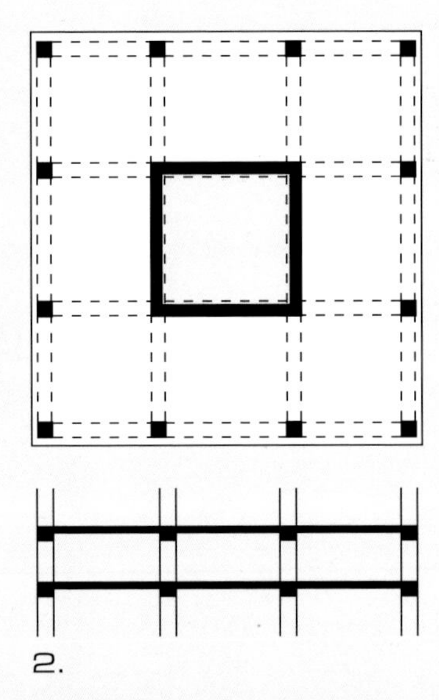

梁和板

梁和板是建筑的横向支撑，它们汇集荷载并将其传送到地下结构。

设计师准确估算建筑结构时（包括材料和减少负重）都必须考虑支柱间的间距，梁覆盖区和板的厚度。

板可以是：

1. 标准厚度板 - 无向的
2. 不同厚度板 - 定向的

LINTEL AND SLAB

Lintel and slab are the horizontal backbone of the building, which have the task of collecting the loads and transferring them to the underlying structures.
The designer, for sizing the structures correctly (regarding material and weight saving) must consider the distance between the pillars, the section of the lintel and the thickness of the slab.

Slabs can be:

1. with constant thickness -undirected
2. with variable thickness - directed

3.

通过分散弯曲形变，在板及梁的组合上可有不同的组合方式，以达到不同的预期效果。

3. 方案A：短梁，厚板

4. 方案B：长梁，薄板

⟷ 荷载支撑方向

By differentiating the warping,
it is possible to obtain different
combinations of slabs and lintels
according to the result to be achieved.

3. solution A: shorter lintel,
thicker slab

4. solution B: longer lintel, thinner slab

⟷ direction of the load support

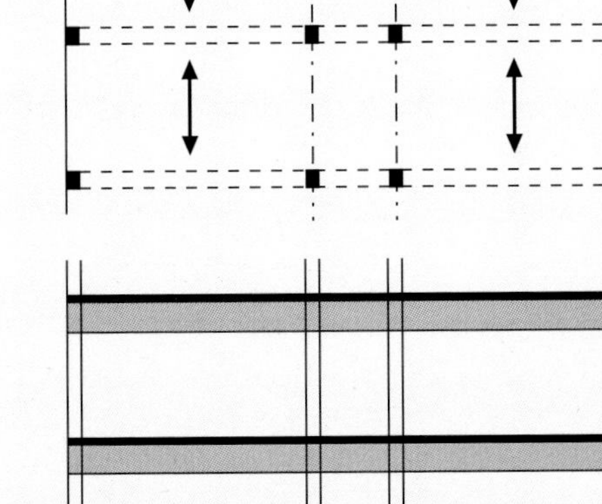

4.

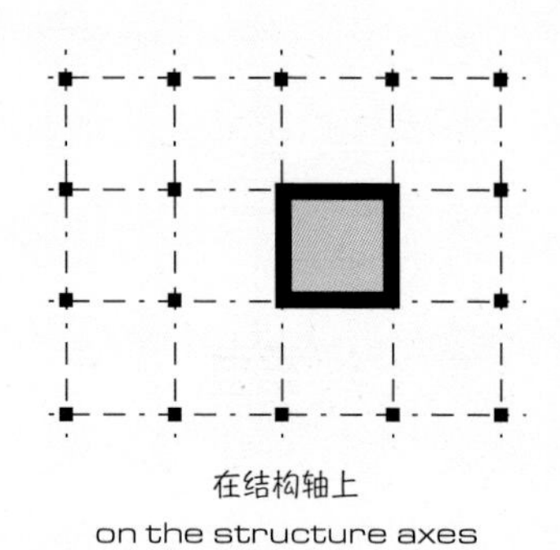

在结构轴上
on the structure axes

开口处

板建造的最关键点在于开口处的选择，如梯井、电梯及其他设施的预留口。

THE PLACING OF OPENINGS

In the construction of a slab, the most critical point of the structure is where openings are inserted, as in stairwells, elevators and slots for service installations.

垂直方向的开口须兼顾构结的支撑，正确的开口处的方法有以下三种：

The right placement of openings for vertical connections with respect to the supporting structure can be made in three ways:

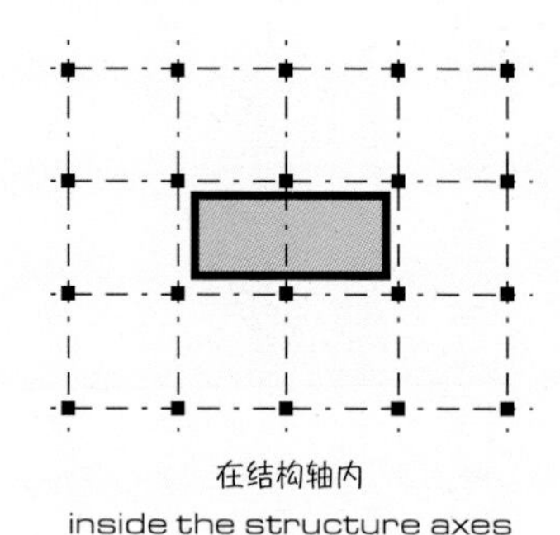

在结构轴内
inside the structure axes

- **在结构的轴向开口**
  嵌入一个空心的承重结构，同时有加固多作用，不会与其他结构冲突（如楼梯，电梯开口）。

- **ON THE STRUCTURE AXES**
  A hollow bearing structure is inserted, which can also have a stiffening function: no conflicts with the structure (typical case is that of stairs and elevators).

- **在结构轴内开口**
  在板中嵌入开口，不更改结构。采用这种方法须特别注意防止破坏承重结构。

- **INSIDE THE STRUCTURE AXES**
  The opening is inserted in a portion of the slab without changing the structure.
  It is important to pay attention and not create any conflicts with the load-bearing structure.

主结构外
externally to the main structure

- **主结构外开口**
  最简便，且最具独立性的方案。

- **EXTERNALLY TO THE MAIN STRUCTURE**
  Easiest and least invasive solution.

施建第一层的过程 PHASES OF CONSTRUCTION UP TO FIRST FLOOR

1 地面
ground

2 去掉腐殖层
humus removal

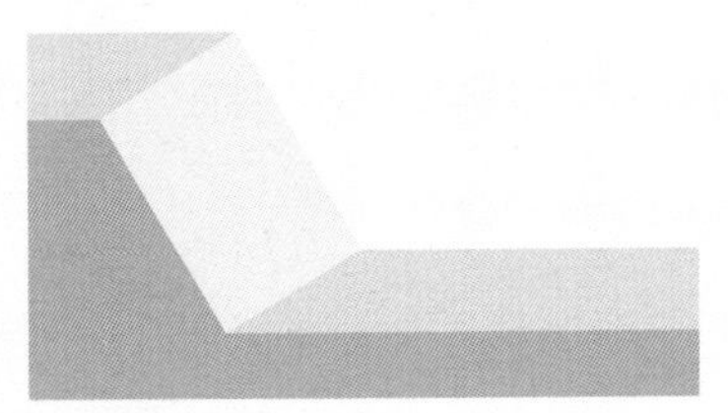

3 一般开挖
general excavation

4 基础开挖
foundation excavation

5 黏合剂
screed

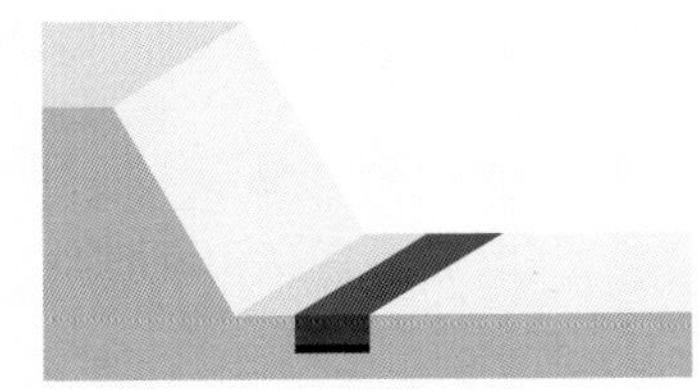

6 基础
foundation

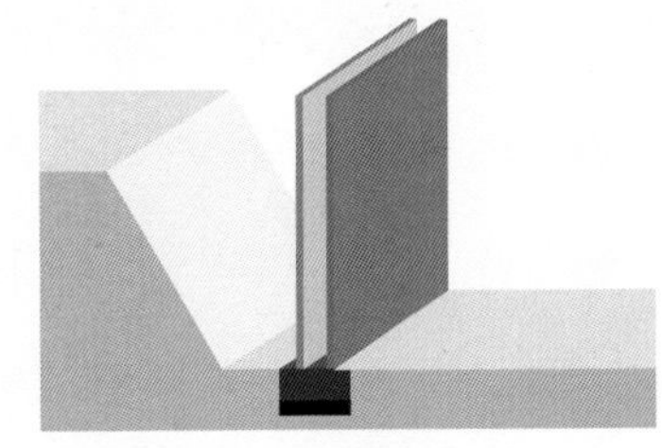

7 模板工程
formwork

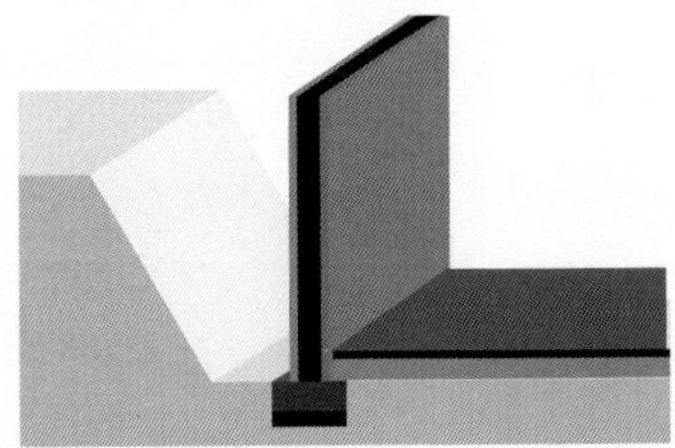

8 浇注墙体和地板
casting of wall and floor

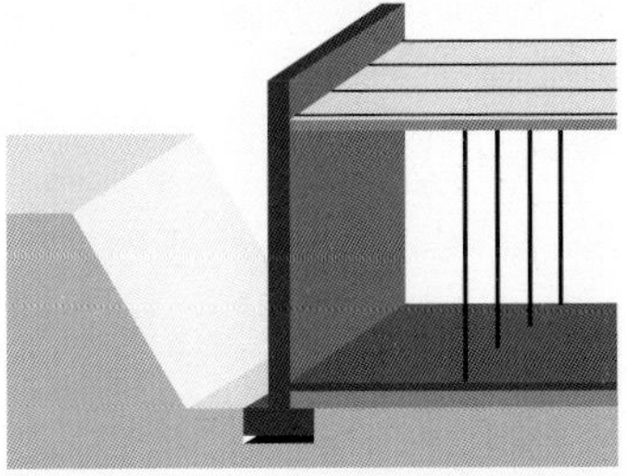

9 模板工程
formwork

10 模板工程
slab

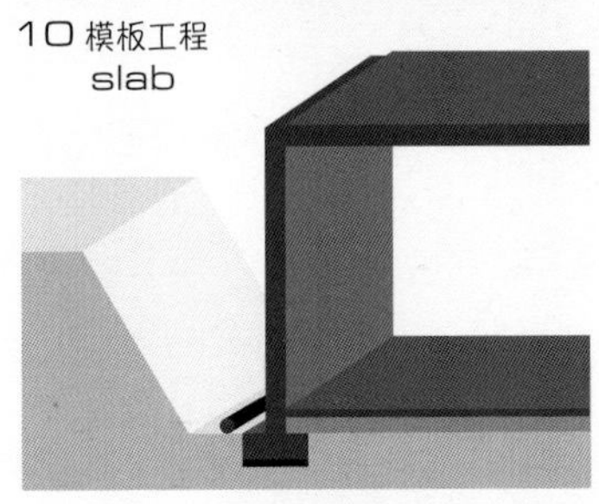

11 排水管道安装
draynage

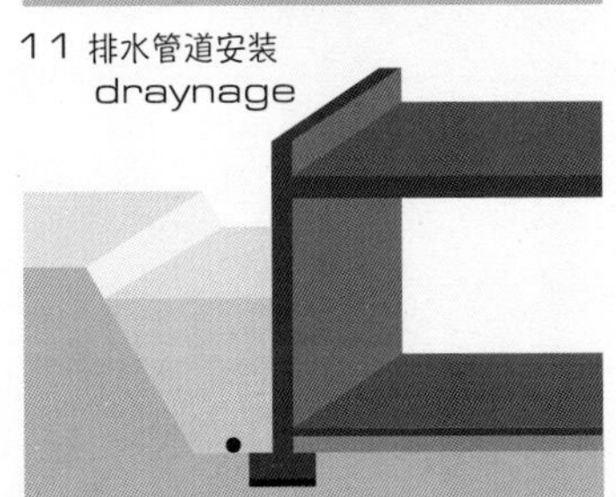

12 填充
filling

法则
PRINCIPLES

结构工程
STRUCTURAL ENGINEERING

建筑外墙
BUILDING SHELL

系统
SYSTEMS

3 建筑外墙

建筑外墙是建筑设计中定义建筑本身及建筑物容积的一个元素。
它既是内、外部的隔离媒介，也是内、外部的联系媒介。同时它还是一个环境元素，定义并呈现外部空间。

每一个建筑外墙须满足一些特殊要求：

3 BUILDING SHELL

The building shell is an architectural element that defines the building and its volume.
Its function is to mediate, separate and connect the inside with the outside, and it is also an environmental element, which defines and identifies the external surrounding spaces.

Each shell must meet specific requirements:

过滤 - Filter	- 界定内部和外部空间 - 内部和外部 - delimitation of the space between inside and outside - 表现形式 - 意像 - 地位标志 - representation - image - status
安全性 - Safety	- 抗撞击 - impact resistance - 防火通道 - behavior in case of fire - 防入侵 - resistance to intrusions
质地 - Wellness	- 水热型 - hygrothermal - 防噪音 - acoustic - 不释放有害物质 - no leakage of harmful substances
实用性 - Usability	- 便于装修装饰 - spaces easy to equip and to furnish - 内部空间利用 - use of the internal space
管理维护 - Management	- 减少能源消耗 - reduction of energy consumption - 经久耐用 - durability - 便于维护保养 - maintenance

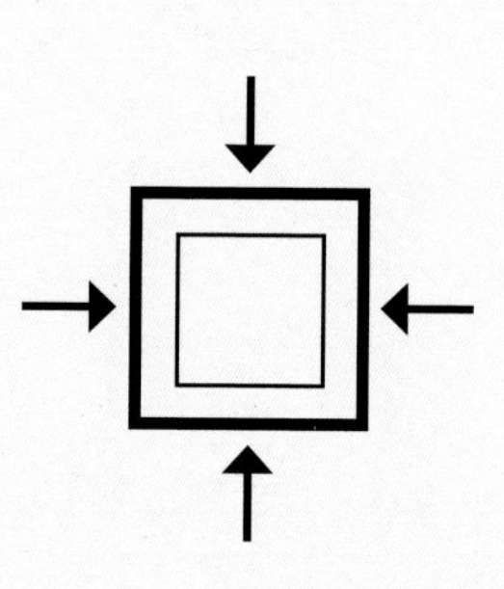

建筑外墙类型

传统的概念忽略了外部墙体、房顶、楼层、天花板与窗户之间的差异而称之为建筑的"皮肤"。
除去功能不说，建筑的外墙就相当于建筑的皮肤。
在这种情况下，我们将建筑的外墙细分为以下几个类别：

BUILDING SHELL TYPES

The classical conception that makes a distinction between external walls, roof, floors, ceilings and windows must be overcome in favor of the concept of "skin".
The building's shell forms the skin of the building regardless of its function.
In this context, we can classify the casing according to specific categories:

重力承受 Resistance to loads	承重 - load-bearing	不承重 - non load-bearing
透光性 Passage of light	透明 - transparent	不透明 - matt
自然通风 Natural ventilation	开口 - opening	不开口 - non-opening
隔热 Thermal insulation	保温隔热 - isolated	不隔热 - non isolated
机械力度 Mechanical strength	强硬 - rigid	灵活 - flexible

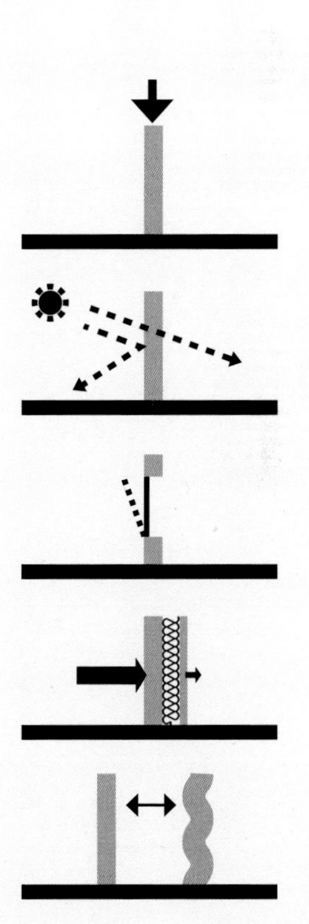

砌墙层

不管是内部还是外部隔板墙都是由不同的材料层组成的。
每层材料都有其特殊功用，与其他材料组合起来能让墙体变得更稳定、持久耐用。
根据不同的要求，砌墙层在厚度、位置和用料上都有不同，甚至没有砌墙层。

STRATIFICATIONS

The partitions, both internal and external,
are implemented by combining different layers.
Each layer performs a precise function and contributes
with the others to provide stability and durability
to the product.
According to the needs, the layers can be modified in
thickness, position, materials or be removed.

铺层 - layer	功能 - function
承重 - load-bearing	- 承担结构本身的重量，以及结构上负担的重量 - supports the weight of the structure and that of the loads above
硬化度 - stiffering	- 抗震及抗悬挂重力与承重层相适应 - confers resistance to shocks or to hanging loads,it often corresponds with the carrier layer
隔离效果 - insulation	- 隔离声音或热量 - thermal and / or acoustic
通风 - ventilation	- 允许水份散发 - allows the dispersion of possible moisture
渗透性 - impermebility	- 密封层（空气，水，蒸气）- sealing layer (air, water, steam)

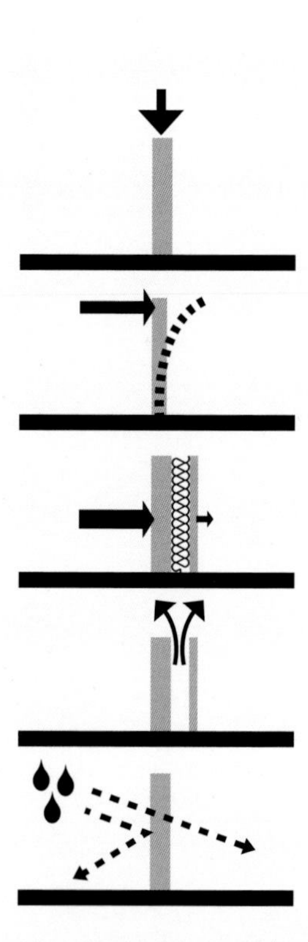

其他问题

在设计和建造的过程中，不论建筑的外墙是何种构成，都需要考虑到以下几个问题：

A 湿度和渗水
B 热桥效应
C 湿度和凝水

A - 渗水

渗水非常危险，会导致外墙体快速毁坏。

ISSUES

Each building shell, regardless of its composition, is subject to problems to which we must pay particular attention in both the design phase and the building phase.
Particularly important are issues related to:

A water infiltrations (moisture)
B thermal bridges (moisture)
C condensation (moisture)

A - WATER INFILTRATION

Water infiltration are dangerous phenomena that lead to the rapid deterioration of the building's shell.

原因 - cause	后果 - effect
- 墙体材料劣质 - low quality of materials - 墙体封口不细致／或防水未做好 - inaccurate realization of the envelope and / or of the waterproofing - 设计失误 - wrong design	- 包裹材料变差腐烂 - degradation and rot of the wrap - 砖体墙盐质外渗 - outcrop of salts on masonry - 给用户的舒适度和健康带来影响 - impact on the comfort and on the health of the user

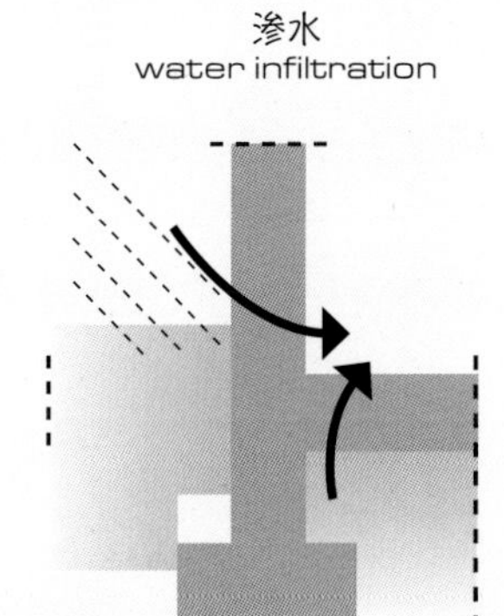

B - 热桥效应

热桥主要出现在以下关键点：

- 柱体与墙体连接处
- 开口处（窗户、门）
- 墙砖与墙体连接处
- 与地面连接处

B - THERMAL BRIDGES

Thermal bridges occur at critical points such as:

- column to shell connections
- openings (windows, doors)
- slab to shell connections
- connections to the ground

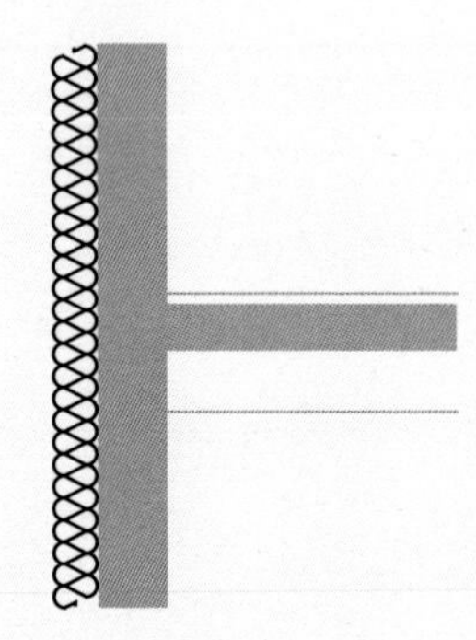

无热桥层，通过外部保护层隔离
absence of thermal bridges: "overcoat" isolation

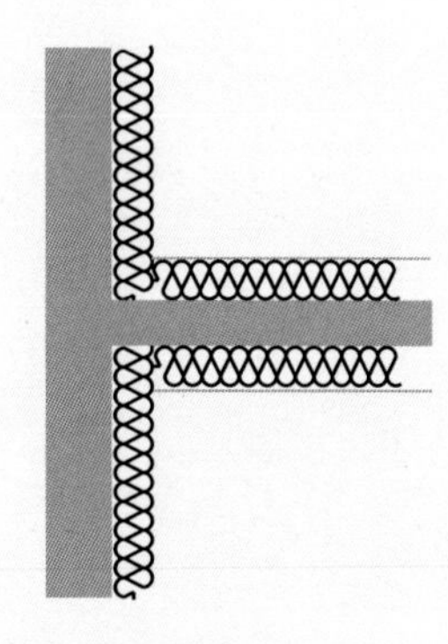

热桥层：连接地板与墙体
thermal bridge: the joint between the floor and the shell

原因 - cause	后果 - effect
- 设计失误 - wrong design - 错误施工 - incorrect production - 无隔离防护 - lack of insulation	- 热量散失 - heat loss - 墙体发霉 - occurrence of mold - 材料腐化 - degradation of materials

C - 湿度和水凝结

内部水汽积聚和墙体水凝形成都是很危险的现象。
这些现象会导致结构快速毁坏，形成有害健康的环境。

C - HUMIDITY AND CONDENSATION

The accumulation of moisture in the interior and the formation of condensation on the walls are dangerous and closely related phenomena. These phenomena can lead to rapid deterioration of the structures and unhealthy environments.

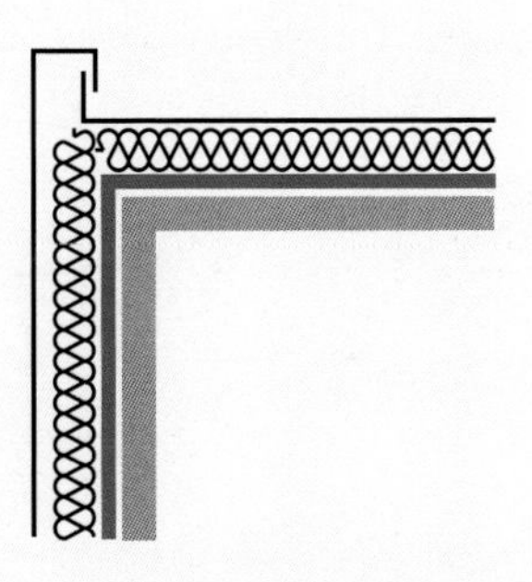

水汽阻隔正确位置
vapor barrier:
correct position

原因 - cause	后果 - effect
- 出现热桥效应 - presence of thermal bridges - 水凝点到达墙体 - dew point is reached on the walls - 缺少水汽阻隔物 - absence of vapor barrier - 室内通风差 - poor ventilation of the rooms	- 房屋腐坏 - degradation and decay of the housing - 开始长霉 - onset of mold - 影响住房舒适感，危害健康 - impact on the comfort and health of the user - 墙体水凝现象 - condensation on the walls

法则
PRINCIPLES

结构和工程
STRUCTURAL ENGINEERING

建筑外墙
BUILDING SHELL

系统
SYSTEMS

4 系统
技术集成系统
空调设施
给排水系统
机械通风和供暖
电力系统

4 systems
technical plants
heating and air conditioning
waterworks and sewerage
mechanical ventilation and heat recovery
electrical system

4 系统 —— 技术集成系统

技术集成系统是建筑物不可缺少的一部分。在当今的设计工作中，系统和建筑物，建筑物和各技术系统都必须综合在一起。因此，建筑技能与技术系统的设计原则须相关一致。

安装建筑内的主要设备可分为：

- 空调设施（供热，制冷，除湿）
- 给排水设施
- 机械通风和供暖
- 电力系统

4 SYSTEMS - TECHNICAL PLANTS

The technical plants are an essential part of the building. In today's design practice, the systems need to be integrated with architecture and the architecture with the plants. For this reason, the skills of the architect must also relate to the principles of system design.
The main equipment installed in a building are classified as:

- heating and air conditioning (hot / cold / humidity)
- waterworks and sewerage
- ventilation system
- electrical system

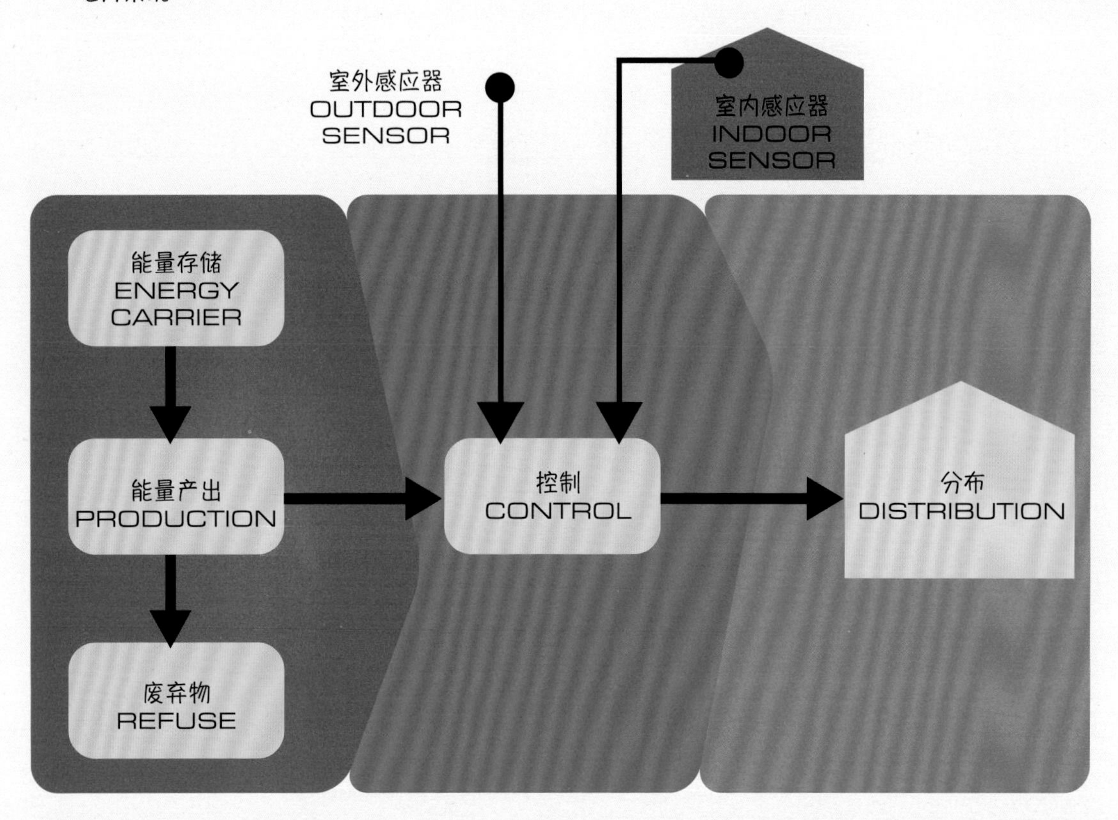

空调设施

另外还有采用不同技术，依赖消耗燃料的商业营利性空调系统。

HEATING AND AIR CONDITIONING

There are different types of commercial heating systems powered with various fuels that offer yields of production in relation to the different technologies used.

COMPARISON OF HEATING SYSTEMS
供热系统对比

输入 INPUT	生产 PRODUCTION	产出 OUTPUT	效率 EFFICIENCY	成本 COSTS *
天然气 GAS	燃烧炉 BOILER	烟 SMOKE	80 %	●●
柴油 DIESEL	燃烧炉 BOILER	烟 SMOKE	80 %	●●●●
生物燃料 BIOMASS	燃烧炉 BOILER	烟 SMOKE	80 %	●●
直流电 DIRECT ELECTRICITY	加热线圈 HEATER COIL	/	100 %	●●●
空气 AIR / ELECTRICITY 电	热力泵 HEAT PUMP	/	300 %	●●
水 WATER / ELECTRICITY 电	热力泵 HEAT PUMP	/	400 %	●●

* Operating costs and depreciation in relation to the lifespan of the system.

给排水系统

给排水系统通过建筑内部的管道系统满足用户不同的用水需求。

WATERWORKS AND SEWERAGE

The waterworks system is responsible for distributing
the different types of water to the users
through a distribution network inside the building.

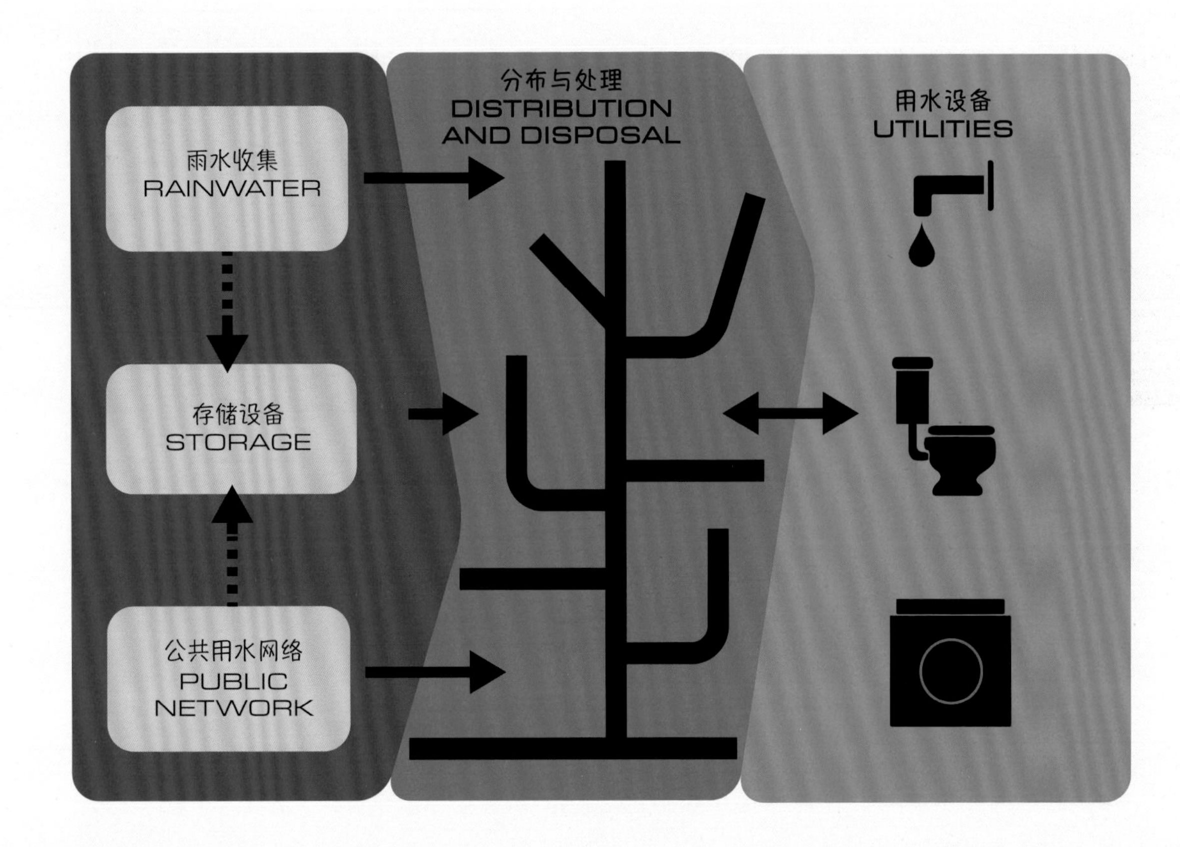

人类产生活动的供水需求通常由公共管道系统提供，灌溉和卫生排污可利用雨水收集系统。

The supply of water for all purposes directly related to human activity usually is from the public network; for irrigation and sanitary discharges it can be through the collection of rainwater.

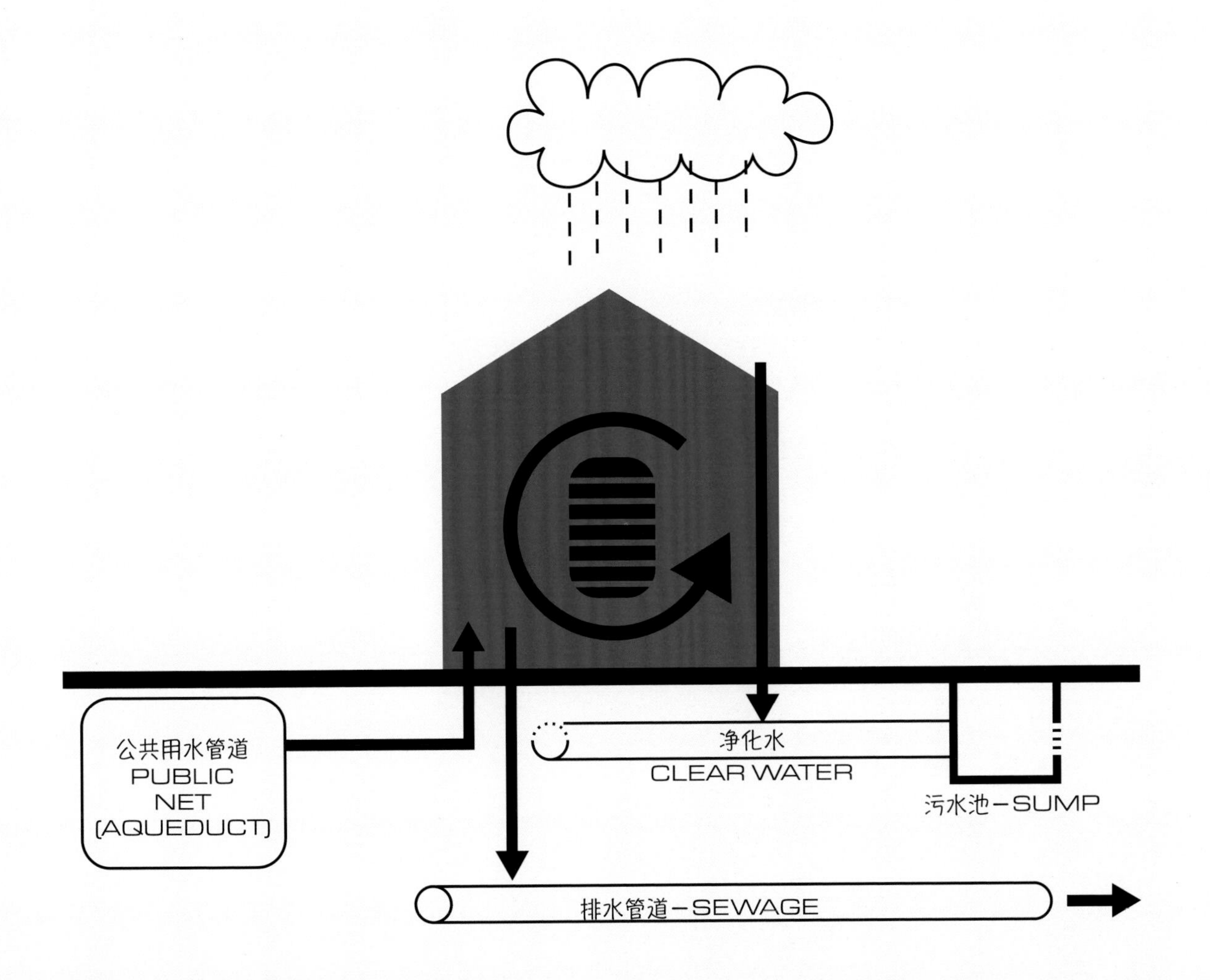

机械通风和供暖

可有效供暖的机械通风系统无须开窗通风，耗能低，不论是在冬天还是其他季节消耗的燃料都少。这种系统已经迅速被各建筑所采用。

地表收集器在冬天可以聚集热量，在夏天可以有效散热。夏天甚至可以不开空调。这套系统对室内舒适度有很大的提升（空气新鲜且富含氧）。

MECHANICAL VENTILATION AND HEAT RECOVERY

The mechanical ventilation system with heat recovery is spreading rapidly in buildings with low energy consumption as it guarantees a significant reduction in fuel consumption in winter and summer, making it unnecessary to open the windows for ventilation of rooms.
The ground collector recovers heat from the ground in the winter and disposes of heat in the summer.
During the summer, it may be unnecessary to use an air conditioner.
The system leads to a significant improvement of indoor comfort (the air is always fresh and oxygenated).

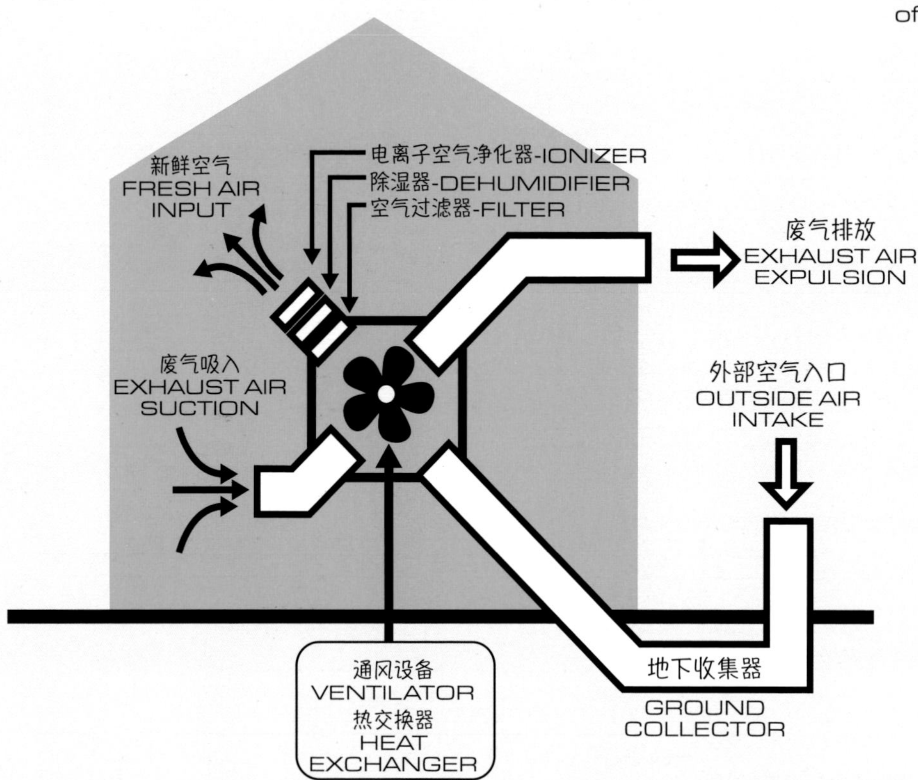

电力系统

电力系统指电力供应和电力设备，主要的功用是传输用电和控制其他系统。
电力来自公共电网或是当地通过再生能源生产的电力(光电板、热电联产，等等)，
之后进入建筑内部电力系统，满足各方面用电需求。

ELECTRICAL SYSTEM

The electrical systems are the electrical supplies and equipment, physical and mechanical, whose main function is to transmit and use electricity or to control other systems.
The energy is taken from the public grid or locally produced through renewable sources (photovoltaic panels, cogeneration, etc.) and enters through a distribution system inside the building to feed the systems that run the building and the electrical outlets.

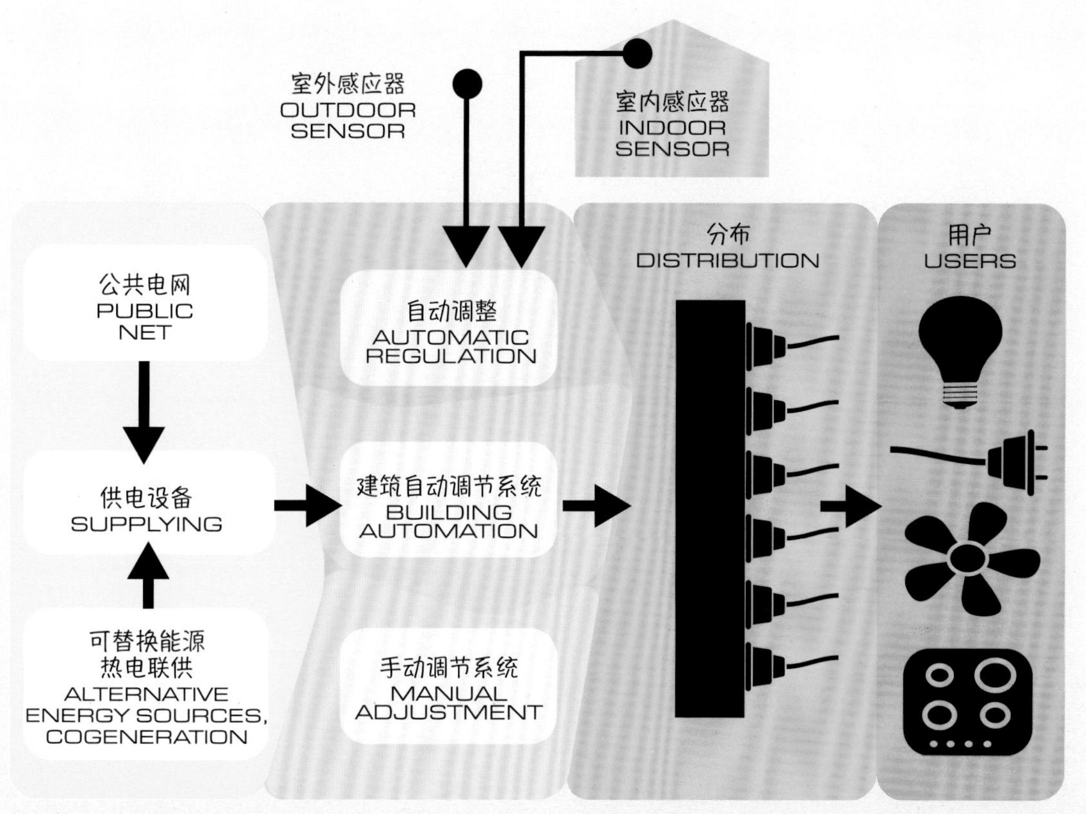

简介
INTRODUCTION

城市与地域
THE CITY AND REGION

未来的城市
THE CITY OF TOMORROW

城市案例
EXAMPLES

1 简介
协调一致性
三大支柱要素
规划者及其责任
空间和思维

1 introduction
coherence
the three pillars
the planner and responsibility
space and mind

1 简介

协调一致性

城市规划者的首要任务就是在设计过程中充分尊重三大支柱要素。
规划者必须将地点、时间、居民的需求三项要素协调一致。
只有尊重这三大基础支柱要素，设计项目才能符合时代要求，充分融入周围环境，人民安居乐业。

1 INTRODUCTION

COHERENCE

The city-planner's task is to respect the three pillars during the design phase.
The planner must be consistent with the place, the time and the needs of the people who inhabit it.
Only based on these three fundamental pillars will he come to a project respectful of the landscape and the environment, in harmony with man and with his time.

三大支柱要素

地点

城市应该是它所在地区的一个产物。
设计方案要与这个地区的特点、地形、气候、环境因素相适宜。

时代

城市必须是入驻文化层次和入住居民的体现，同时，它也应该是"时代的产物"。

用户

城市并不是一个空空如也的容器，生活在其中的人们才是最重要的。他们的需求、兴趣、习惯都必须得到尊重和满足。

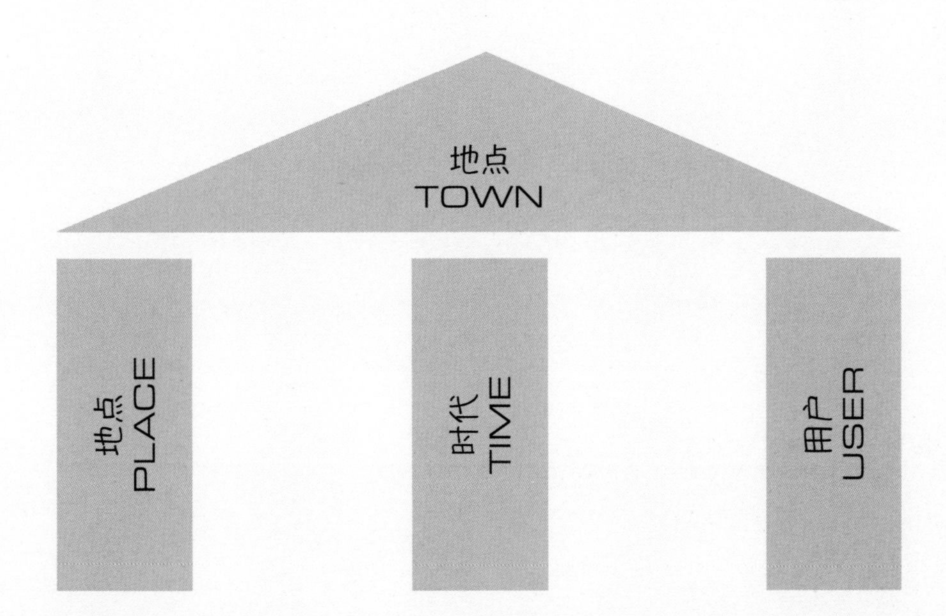

THE THREE PILLARS

PLACE

The city must be the result of the place where it stands.
The project must be adapted to the characteristics of the place, the morphology of the terrain, the climate and its environmental variables.

TIME

The city must represent the stratigraphy of the cultures and peoples that have inhabited it and, at the same time, it must be "a child of its time".

USER

The city is not simply an empty container, but is made up mostly by the people who live in it.
It must meet the needs, interests, habits of those who enjoy it.

规划者及其责任

规划者肩负着重大的责任，每一项设计的选择都会给社区带来实际的影响。
起草设计项目的规划者是担负直接责任的第一人。他对以下几项负有直接责任：

- 社区
  对社区居民，以及公共空间的利用者在共同利益、福利、安全方面的影响负有责任。

- 客户
  负责项目的圆满完成，负责项目的质量及耐用性。对经济成本负责。

- 公共管理
  技术方面必须满足城市或建筑大楼的各项参数要求。

THE PLANNER AND RESPONSIBILITY

The planner has a role of great responsibility:
every design choice falls in fact on the community.
When the planner draws up a project, he is directly responsible to:

- the community
  responsibility for the impact on the territory (the common good), the welfare and the safety of the occupants and users of public spaces.

- the customer
  responsibility for the success and the quality and durability of the project;
  responsibilities for economic / finance.

- public administration
  technical responsibility for compliance with the urban / architectural codes.

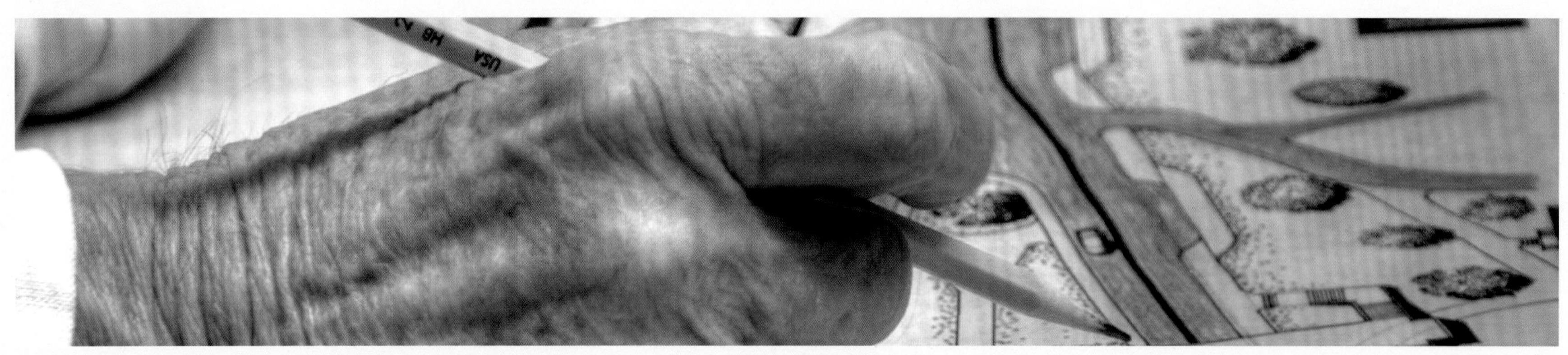

123

空间和思维

建筑和城市的规划设计对人们有着重大的影响，这种影响从来都不是中性的，它既可能是正面的，也可能是负面的；既可能丰富我们的生活，也有可能毁坏我们的生活。

我们居住的空间会影响我们的心情、我们的思想和行为，以及和别人交往的方式。

我们的视觉、听觉、嗅觉对我们的心理都有非常大的影响。

明亮的颜色、光线、自然景观，以及和谐的声音会让我们情绪积极亢奋；相反，暗色调、光线昏暗、空间逼仄、糟糕沉重的人文风貌让我们阴郁呆板。

反过来过程也一样。
在设计过程中，设计师的情绪都倾注到了产品中，产品也因这位设计师的心理和情绪而有了各自不同的蕴含意义。
正如我们开辟空间一样，空间也塑造着我们的思想。

SPACE AND MIND

The architecture and urban planning have a great influence on people, both positive and negative; enriching or impoverishing our lives deeply, they are never neutral.

The space in which we live can affect our mood and therefore our thoughts, our behavior and the way we interact with others.

The set of visual, auditory and olfactory perceptions have a strong influence on our psyche.

The use of bright colors, sunlight, natural landscapes and harmonious sounds have a positive influence on our mood; on the contrary dark colors, lack of light or space, and degraded or heavily anthropic landscapes have a negative influence on our thoughts.

But this same process also works in reverse.
During the design phase, the mood of the designer is poured into the product, giving the product different connotations depending on the psyche and emotions of the designer.
As we create the space, the space in which we live shapes our minds.

124

2 城市及领域
超级城市和城市体系
城市－功能
城市－三个层面的分析
城市－网络和结构
城市－构成元素

2 city and territory
hyper-city and systems of cities
city - the functions
city - three levels of analysis
city - networks and structures
city - constituent elements

2 城市及领域

城市设计就是处理景观主题和地势地形。

城市所在的地区影响其构造和发展。
因此，城市地形与公路构成以及与周边的入口设定都有着紧密联系。

对城市公路网络的分析可推测出一些重要信息，如它的历史起源、将来的发展和运作。

通过分析和比较城市地形与它所处地区的关系，我们总结出5种类型的城市规划：

城市的类型

- 辐射型城市
- 坐标型或直线型城市
- 网格型城市
- 全面都市化
- 非都市化地区

2 CITY AND TERRITORY

Designing the city is dealing with the themes of landscape and territory.
The place where the city stands strongly affects its structure and growth.
There is thus a close relationship between the morphology of the city, the structure of its roads and access from the surrounding area.
From the analysis of the road network of a city, it is possible to deduce important information about its origins, possible development and operation.
By analyzing and comparing the morphology of cities in relation to the territory in which they arise, we can distinguish 5 main types of urban plots with which to classify the city, its territory and the surrounding areas:

TYPOLOGIES OF CITIES

- radial city

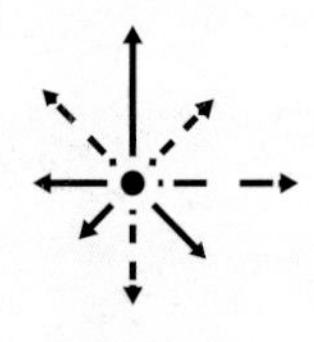

- axial or linear city
- grid city
- widespread urbanization

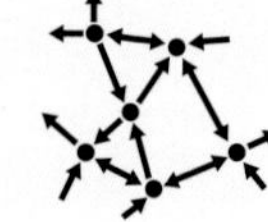

- non-urbanized territory

辐射型城市

辐射型城市的特点是有一个原始城市核心区，从核心区分多支向外辐射。

优点：中心地区和外围直线联系

缺点：中心地区交通拥堵

RADIAL CITY

The radial city is characterized by the presence of an original nucleus, from which axes of communication branch radially.

positive: recognizable connection axes between center and periphery

negative: traffic problems in the center

MILANO
米兰

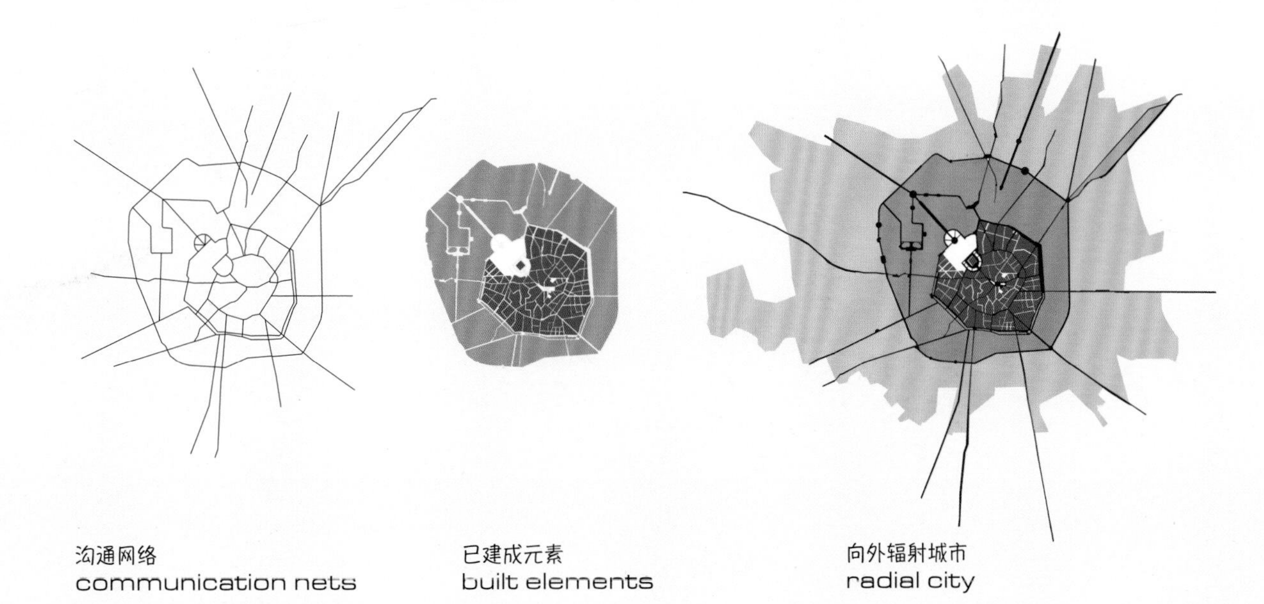

沟通网络
communication nets

已建成元素
built elements

向外辐射城市
radial city

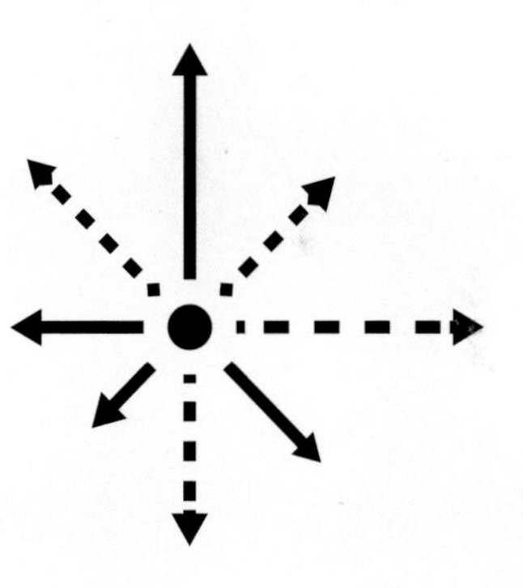

坐标型或直线型城市

坐标型或直线型城市的特点是有一条主线路，沿线地区随之开发起来，从而又衍生出二级公路系统。

优点：坐标沿线都有发展机会

缺点：交通距离长

AXIAL OR LINEAR CITY

The axial or linear city is characterized by the presence of a main route along which develops the urban territory and from which originates the secondary road system.

positive: opportunities for growth along the axes

negative: long distances

BRASILIA
巴西利亚

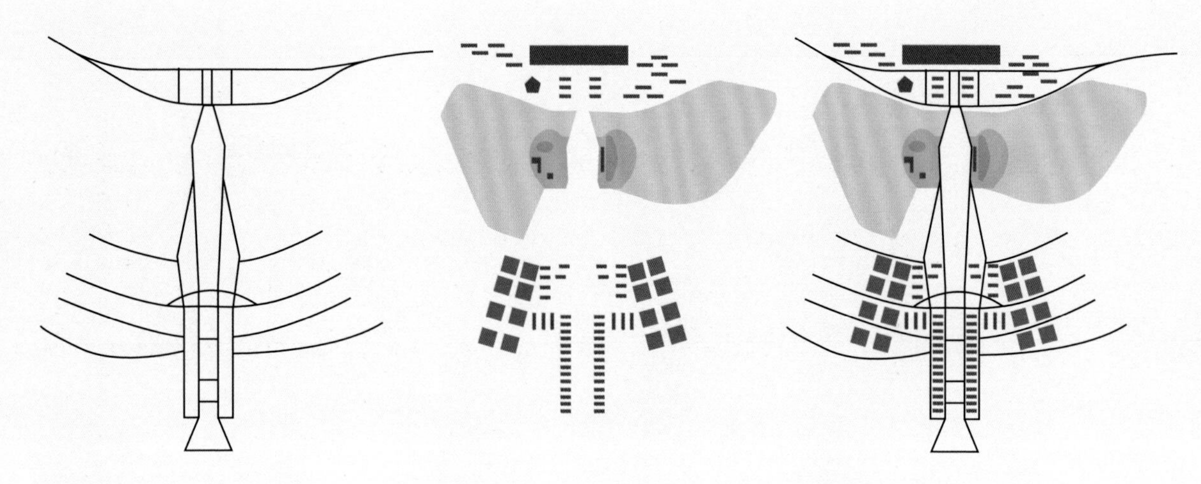

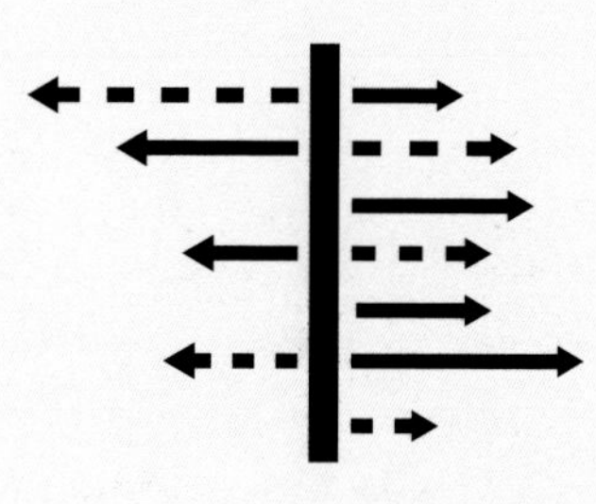

沟通网络
communication nets

已建成元素
built elements

轴线城市
axial city

网络型城市

网络型城市的主要特点是公路网络看起来像一个规整的网格。网格的节点处分布着一些重要地点。

网格的中间分布着重要的元素。

优点：清晰有序

缺点：没有可作为参考的中心点

GRID CITY

The grid city is characterized by the presence of a more or less regular grid, which defines the texture. A number of places of interest coexist, located in the nodal points of the grid.
The elements are positioned, and so directed, along the meshes of the grid that constitute the backbone of the network of connections.

positive: order and clarity

negative: no central point of reference

CHICAGO
芝加哥

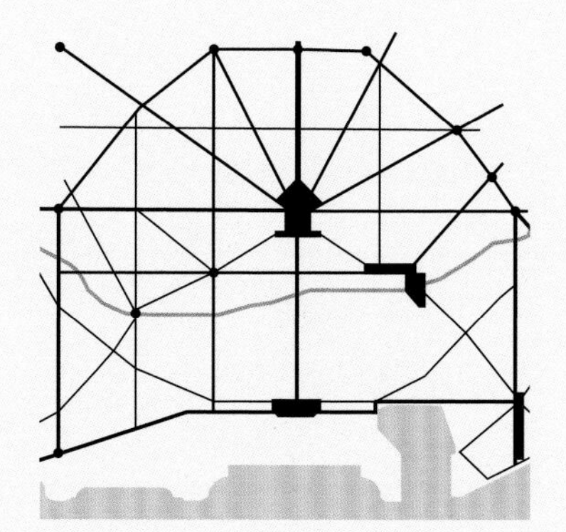

沟通网络
communication nets

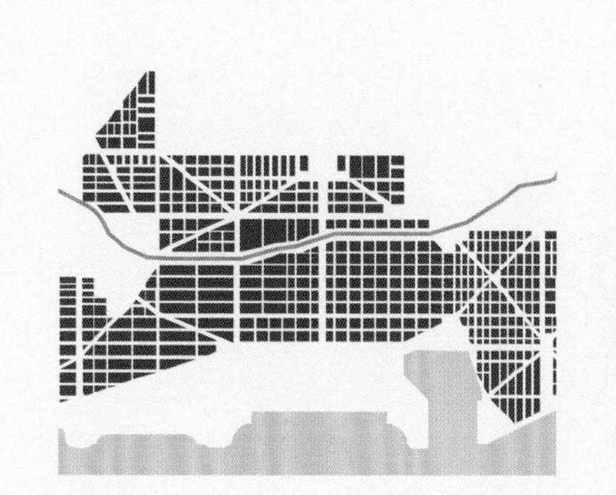

已建成元素
built elements

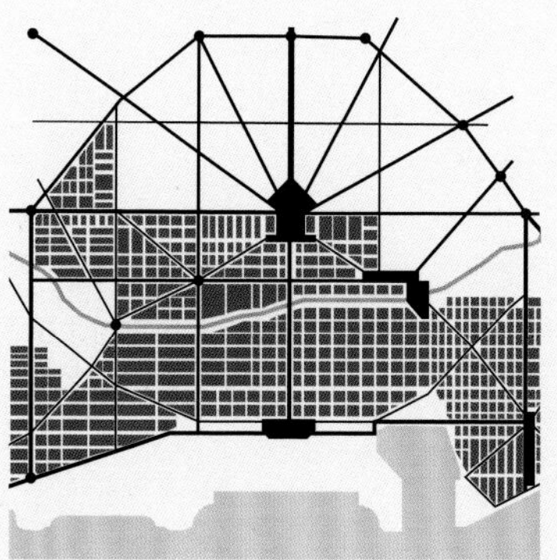

网格型城市
grid city

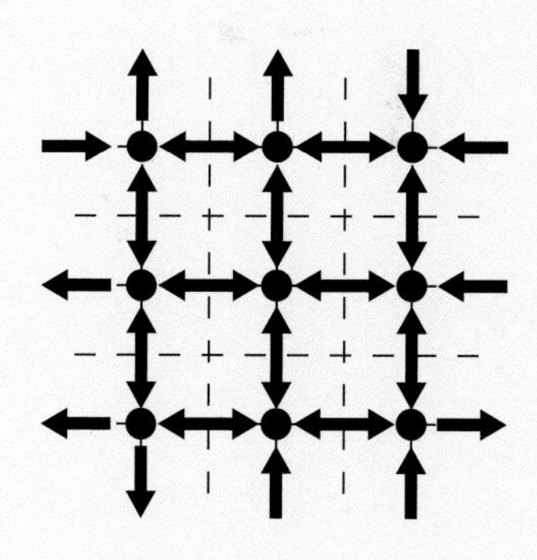

全面都市化

整个城市在都市化范围内，没有特定的界线，实现基础功能的不同城市元素不规则地分布在城市范围内。

优点：极大的个体自由

缺点：远距离交通
　　　需要依靠小汽车
　　　无社交活动
　　　无界线
　　　人口密度低（土地消耗量高）

URBANIZATION WIDESPREAD

The sprawling city is a portion of the urbanized area characterized by the absence of a defined limit and the presence of disparate elements with nodal basis functions, arranged in no particular order that is recognizable within the territory.

positive: great individual freedom

negative: long distances
dependence on cars
no social involvement
lack of boundaries
low-density (high consumption of soil)

BRIANZA

布里安扎

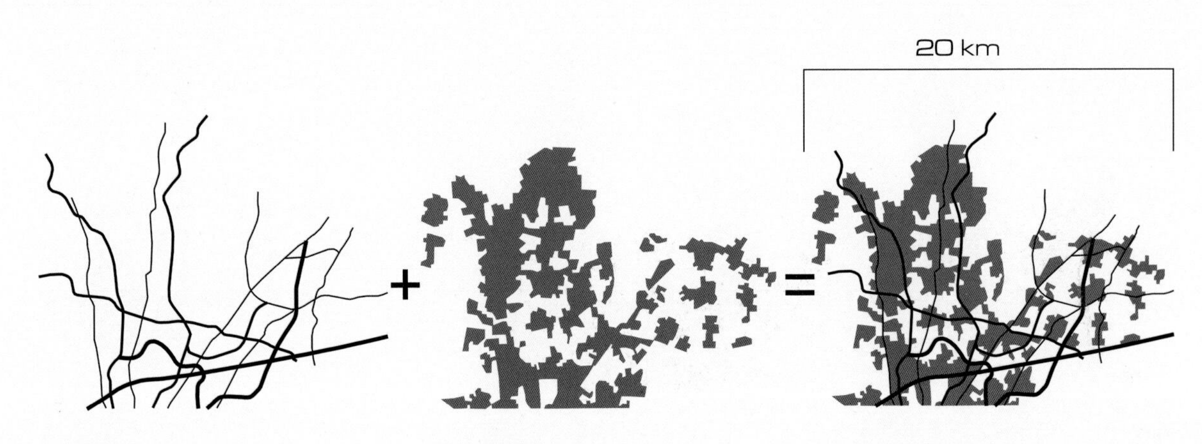

沟通网络
communication nets

已建成元素
built elements

城市蔓延
urban sprawl

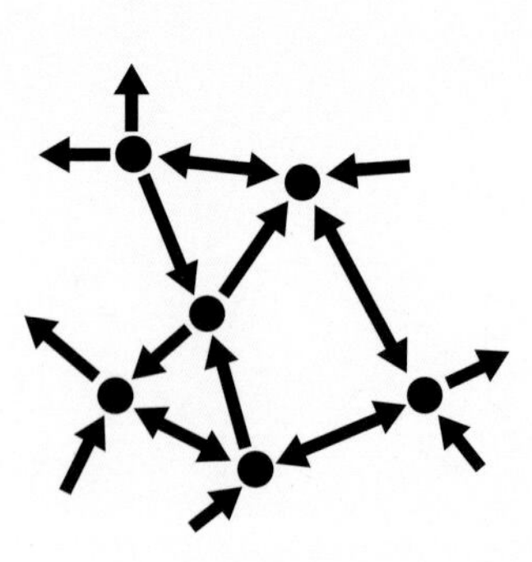

未都市化地区

未都市化区域人口密度低，中心居住区很少，(甚至没有) 有大片的绿林区。

优点：人口密度低

缺点：远距离交通
社交机会少
需要依赖汽车

NON-URBANIZED AREA

The non-urbanized area consists of low density urban territory, characterized by the presence of very few inhabited centers (or their total absence) and large green areas available.

positive: low-density

negative: long distances
lack of opportunities for social interaction
dependence on cars

AREA OF LOWER LOMBARDY
伦巴第城区

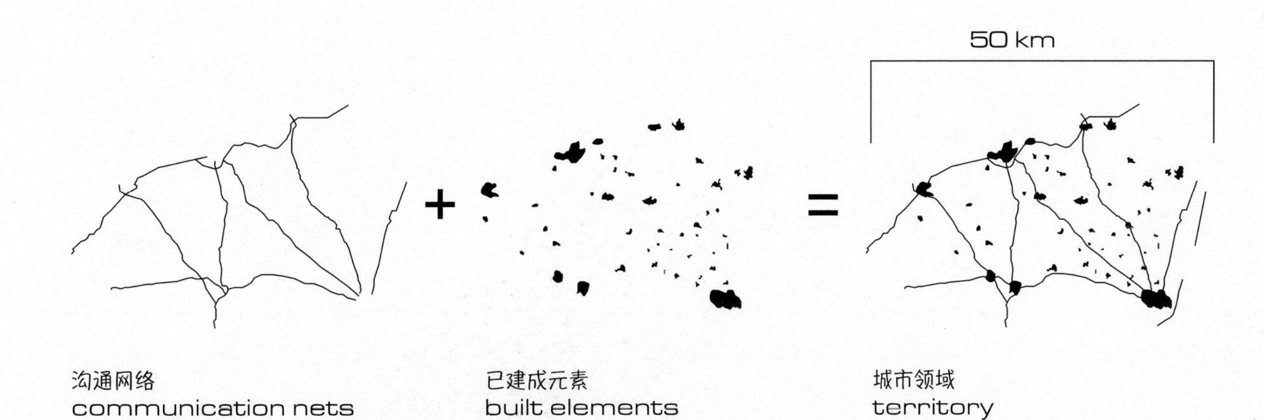

沟通网络
communication nets

已建成元素
built elements

城市领域
territory

超级城市和城市系统

HYPER-CITY AND SYSTEMS OF CITIES

大都市

METROPOLIS

大都市是指市区居民人数达到，甚至超过了一百万的大城市。该城市是地区或国家的政治文化中心，是各项交流的中心，是周边地区的引力极点。

A metropolis is a large city whose metropolitan area reaches or exceeds one million inhabitants.
Economic and cultural center of a region or a country, the metropolis is the nodal center of the communication flow and an attraction pole for the surrounding area.

优点：极大的个体自由
人口众多

缺点：社交活动与距离成反比，离人口密集的中心区越远，社交越少。

positive: great individual freedom
large population

negative: phenomena of commuting
social involvement is inversely proportional to the distance from the core
high population density in the central districts

伦敦
伦敦是欧洲人口最稠密的大都市，居民人数将近830万，人口密度达到了5200人/平方公里，城市面积1570平方公里。

LONDON
The City of London is the most populous and widespread European metropolis with a population of approximately 8300000 inhabitants, a settlement density of 5200/km^2, a territorial extension of 1570 km^2.

超级大都市

超级大都市是居民人数超过2千万的大面积城市。

超级大都市是由几个大都市组成的一定地区范围的独特城市体。

优点：极大个体自由
人口众多

缺点：远距离交通
需要依赖汽车
社交活动少
土地消耗量大

MEGALOPOLIS

The megalopolis is an urbanized area very large (at least 20 million inhabitants).
It has a regional dimension and is characterized by the presence of several smaller metropolitan areas gathered to create a unique body.

positive: great individual freedom
large population

negative: long distances
dependence on automobiles
low social involvement
high consumption of soil

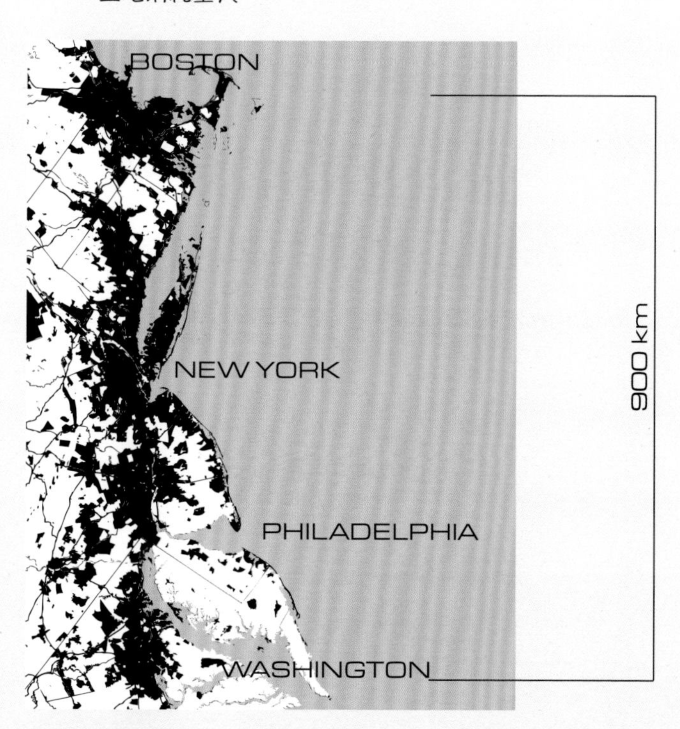

大西洋都市带
（波士顿-纽约-费城-华盛顿）
超级大都市沿美国东海岸，从波士顿一直延伸到华盛顿，被大家称之为“波华城”。该地区长900千米，居民人数5200万。

ATLANTIC MEGALOPOLIS
(boston - new york - philadelphia - washington)
The megalopolis stretching from Boston to Washington (commonly known as BosWash) runs along the east coast of the USA.
It grows to a length of about 900 kilometers and has a population of 52 million inhabitants.

生活
live

工作
work

休闲
recreate

城市 —— 功能

城市是较稳定的人类居住区，城市周边则常住人口密度相比较低，服务设施和基础建设也相比较少。

城市由居民区、商业区、管理行政区、工业区及人们可以相互交流的休闲娱乐中心区组成。休闲时间人们能相互交流凝成一个整体。

城市里的活动主要有以下几项：

- 生活
- 工作
- 休闲

城市的主要功能就是为居民创造最好的生活条件。

CITY - THE FUNCTIONS

A city is a stable human settlement that differs from the surrounding area due to its population density and the presence of services and infrastructure.
The city consists of residential areas, commercial areas, administrative areas, industrial areas and areas for recreation and leisure time interacting with each other to form a single body.

The main activities that take place within a city are:

- live
- work
- recreate

The main function of a city is to create the best possible framework conditions for the life of its inhabitants.

城市 —— 三个层面的分析

从宏观到微观的顺序，城市的组成元素有以下几种：

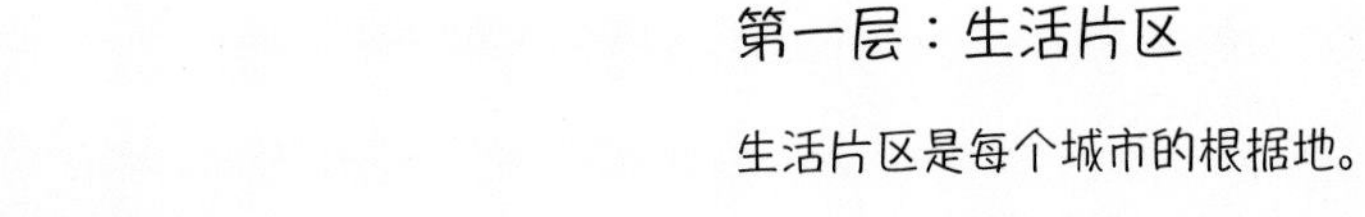

第一层：生活片区

生活片区是每个城市的根据地。

生活片区能满足人们开展各项社会活动的功能需求，同时建立一个集会和商业中心。

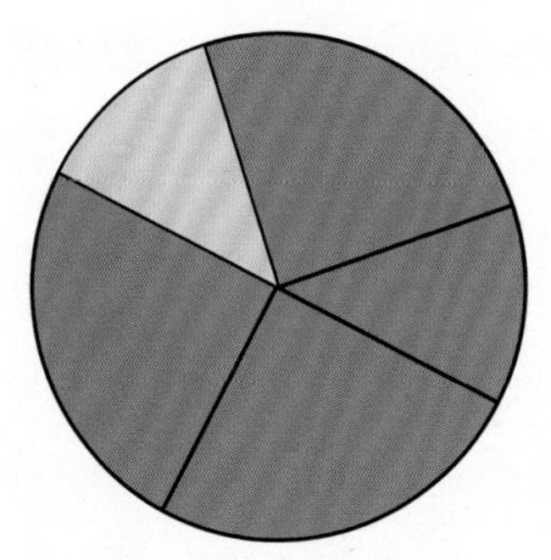

生活片区的主要特点有：

- 服务齐全
- 各种功能目的地的集合
- 引力地点的集合（邮局、加油站等）
- 良好的内部和外部沟通
- 安全保护到位
- 片区定位
- 特色识别性

CITY - THREE LEVELS OF ANALYSIS

The constituent elements
that make up the city
(from macro to micro) are:

LEVEL 1 - DISTRICT

The districts are the characteristic anchors of each city.
The district should provide all the essential functions for the conduct of social life and at the same time establish a center for meeting and commercial attraction.

The district, therefore,
is characterized by:

- presence of services
- mix of destinations
- presence of attractors (post office, gas station, ...)
- good internal and external communication
- protection from hazards
- position
- identity

第二层：街道，广场，街区

街道、广场和街区就像砖块一样，是组成城市的重要元素。
公路网络与街区直接联动，形成一个城市网格，也是城市的结构。

街道、广场和建筑必须：

- 保证高品质生活
- 各种功能区可以组合搭配（居住、行政、商业、运动、休闲和文化）
- 有序分流各种不同交通（快 - 慢、机动车 - 行人、内部 - 外部）
- 按不同的速度规划交通
- 街道和广场不仅仅是交通地点

LEVEL 2 - STREETS, SQUARES, BLOCKS

Streets, squares and blocks are the bricks, the constitutive elements of the city.
The network of roads directly interacts with the blocks to form the city grid and the city structure.

Streets, squares and buildings must:

- ensure a high quality of life
- allow a balanced mix of use (residential, administrative, commercial, sport, leisure and culture)
- hierarchically separate the different types of traffic (fast - slow; vehicular - pedestrian, internal - external)
- shape traffic depending on its speed
- not only be places of traffic

第三层：建筑和空间

分析一个城市“虚与实”的关系，即建筑实体所占空间与开放空间之间的关系有着非常重要的意义。

一般情况下，规划者容易倾向于分析实体建筑，然而实际上建筑之间的空间和空间质量更重要。

开放空间组合成一个标志性空间，人群在其间流动，自发形成行使城市功能的活动中心以及人们社交活动的中心。

LEVEL 3 - BUILDINGS AND SPACES

Of fundamental importance in the analysis of a city is the relationship between the “full and empty", between the space occupied by buildings and open spaces. Normally planners tend to analyze the full (buildings) more readily, while the gaps and their quality are more important. The voids combine to create identity, flows of people and dynamism within the city, as well as to identify the centers around which to articulate the main functions of the city life and social interaction among people.

城市——网络与结构

城市地区分布着不同的网络结构，这些网络相互交叉形成了建筑空间与开放空间的特殊关系，从而也形成了每个城市独有的特点。

CITY - NETWORKS AND STRUCTURES

Within an urban area, there is a sequence of networks that intersect with one another and, through the special relationship between the built volumes and empty spaces, contribute to the identity of each city.

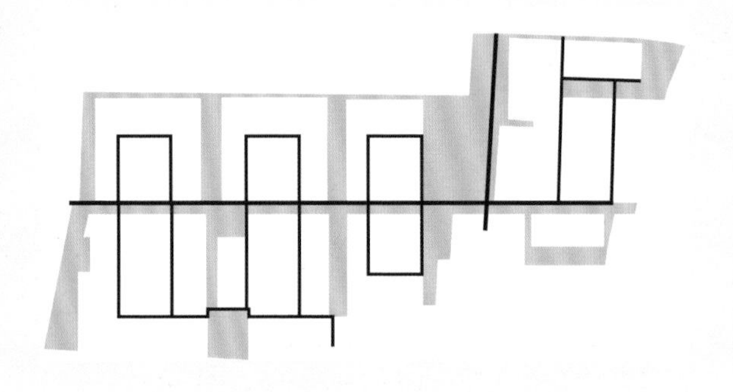

绿地网络

绿地空间是城市的互动构成元素。

GREEN SPACE NETWORK

It is the set of green spaces that interact with the city.

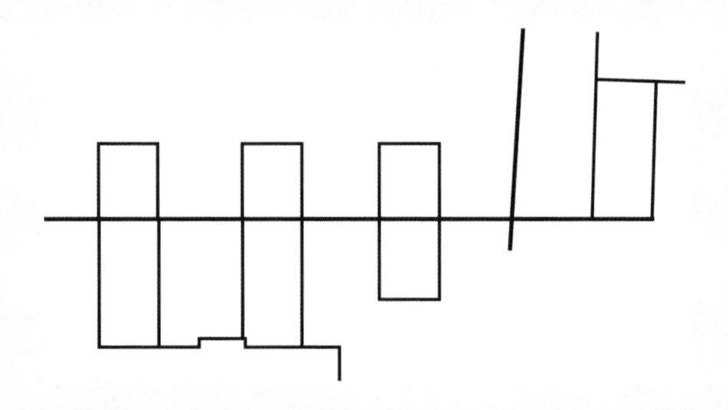

运输网络

清晰标示城市的特点和交通流向。

TRANSPORT NETWORK

It helps to delineate the character of a city and identifies the traffic flows.

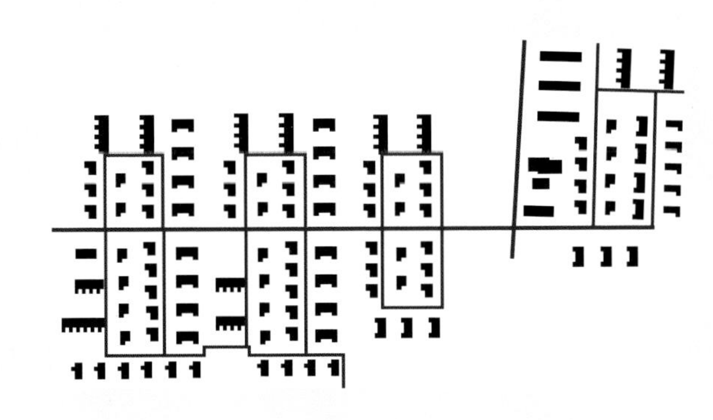

容积和开放空间网络

建筑容积也是网络的组成部分：

- 建筑容积网络（实）
- 开放空间网络（虚）

NETWORK OF VOLUMES AND SPACES

Even the volumes constitute networks:

- network of the built volumes (full)
- network of the open spaces (voids)

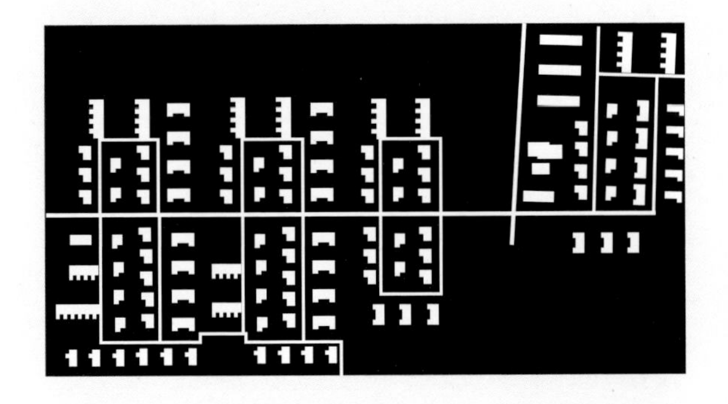

服务网络

21世纪，服务网络将是城市构成的重要基石：

- 健康与社交服务网络
- 商务服务网络（商业、工业、农业）
- 技术服务网络（WiFi网络、光缆网络）

SERVICE NETWORK

In the twenty-first century, they are a fundamental cornerstone for the city:

- network of health and social services
- network of services for business (commercial, industry, agriculture)
- network of technological services (WiFi network, cable networks)

哥本哈根的都市延伸项目

Bäuerle-Lüttin Architekten

Urban expansion project in the city of Kippenheim (D)

Bäuerle-Lüttin Architekten

城市——构成元素

一个人类居住区，不管其量化特征如何，只要它有自己的实质精神就能成为一个城市。

灵魂

一座名副其实的城市应该具有：

- 特色标志
- 城市灵魂

城市的标志来自于它独有的构成元素和与周围地区不同的特点：

- 有明确的城市界线
- 有城市的出入口、港口

城市的灵魂来自于赋予城市生命力和特点的元素，且区别于其他城市的构成元素：

- 地标建筑
- 实物建筑与开放空间的关系
- 通道（方便人们社交互动）
- 保存城市居民历史和文化形态

CITY - CONSTITUENT ELEMENTS

A human settlement, regardless of the quantitative characteristics, fully deserves the status of a city if it has an own qualifying spirit.

SOUL

The city worthy to the name should have:

- a characteristic identity
- its own soul

The identity of a city must emerge through the elements that distinguish it from the surrounding area and giving its independence:

- a clear boundary as an element of separation
- ports such as access points

The soul of a city is formed by the elements that give it life and characterize it, making it different from other cities:

- landmarks and reference sites
- the relationship between solids and voids
- paths (for socializing and interaction between people)
- morphology in keeping with the history and culture of its inhabitants

边界线

边界线将城市与周边地区区别开来，形成城市区与非城市区。

边界线表示归属身份关系，明确城市里的内含元素和城市外的外在元素。

现代城市有两类边界线：

- 明确的边界线
  原始城市与乡村的区别通常以实物元素为标志，如建筑体（城墙）或是自然实物（河流）

- 未明确的界线
  典型的例子是不规则的城市扩张，城市领土以不确定的方式向外扩张，同时形成了一些城市边缘地带

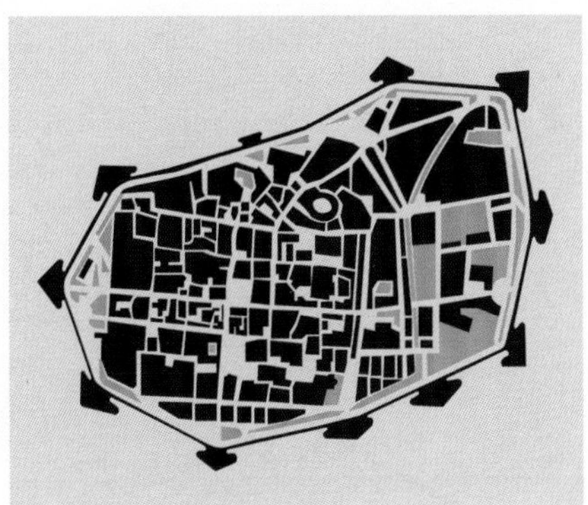

卢卡（意大利）
Lucca (italy)

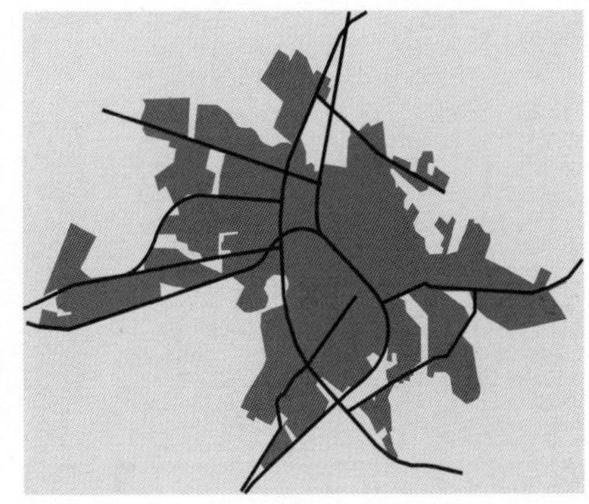

阿雷佐（意大利）
Arezzo (italy)

THE BORDER

The border is the element that separates the city from the surrounding area, which creates the city, separating it from what is not the city.
The border expresses a concept of belonging and identity, identifying a line that establishes relationships of inclusion and exclusion among the elements inside and outside the city.
In the contemporary city, we distinguish between two types of boundaries:

- the precise, defined boundary: typical situation of the cities of ancient formation in which the separation between city and country is clear and identifies with a physical element, built (city walls) or natural (river)

- the undefined boundary: typical situation in unregulated urban expansion, which creates urb an fringes that creep in a chaotic manner in the territory

入口门户

从历史上来讲，港口的概念就是进入城市的门户，而进入门户就意味着穿越界线进入城市。

港口主要有以下功用：

- 标志城市的入口
- 调节出入城市的流量
- 标志城市的魅力元素

具有历史意义的城市港口极具特色，是城市生活的主要构成。

THE ENTRANCE DOOR

Historically, the concept of port as the entrance door to the city is closely linked to that of the border because the city gates represent the crossing points across the city border.

The port has the function of:

- identifying entry points to the city
- adjusting the flow passage to and from the city
- presenting themselves as attractive element of recognition

In the historic city, doors were easily recognizable and assumed a major role in city life.

125

加迪斯 - 西班牙
Cadiz - Spain
Puerta Tierra Cádiz

过去，城市的门户具有一定的特色，极易识别，而如今这种概念不再普遍适用。

港口门户的概念与进入城市相关，而不再是原来穿过城墙进入城市的概念。
门户的功能属性与现代交通元素相关，如：

- 火车站
- 机场
- 公路收费站
- 海港和内陆港

While in the past the gate of the city was easily recognizable and identifiable by certain characteristics, today this concept is no longer generally applicable.

The concept of port is connected to the point of arrival in the city that no longer corresponds with the gap to be crossed in the walls.
The function of the door falls on other elements of a different nature related to modern mobility, such as:

- railway stations
- airports
- tollbooths
- seaports or inland ports

阿姆斯特丹（荷兰）——交通运输网络及入口
Amsterdam (The Netherlands) - the transport nets and access

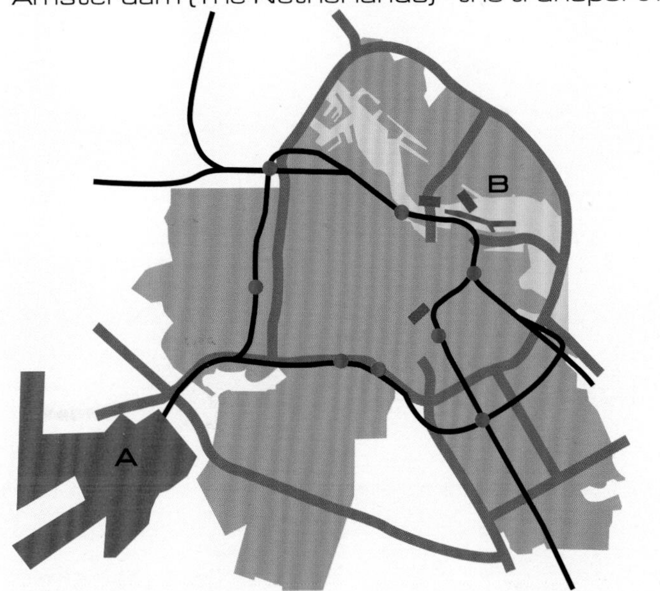

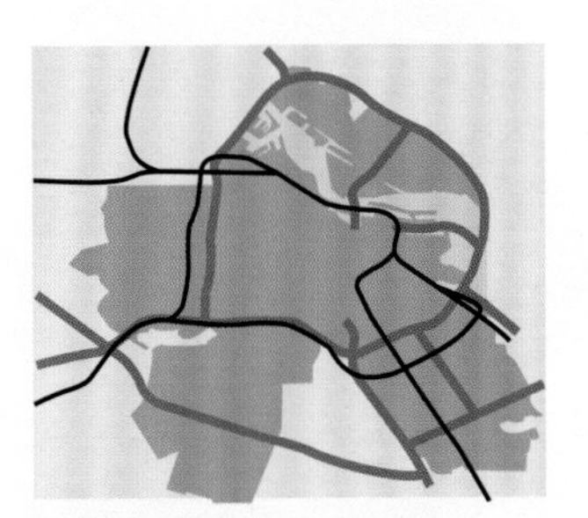

阿姆斯特丹 - 荷兰
Amsterdam - NL

铁路线及车站 - rail nets and stations

高速路和收费站 - motorways and toll

A 机场 - airport

B 海运 - marina

名胜古迹

名胜古迹是那些历经时世变迁而留下的或可转移的，能吸引人群的地方。
它们的作用就像是城市交通网络里的节点媒介，可以引导人流量，并构成节点与节点之间的联系。

名胜古迹可以是：

1. 历史纪念碑
2. 圣地
3. 文化中心（博物馆、画廊等）
4. 休闲中心（运动馆、饭店等）
5. 广场和开放空间

POINTS OF INTEREST

The points of interest are those fixed or movable over time, generating flows of people and attraction.
They function as catalysts nodes of a network of mobility within the city by moving people and relationships from one node to another.

Points of interest can be:

1. historical monuments
2. places of worship
3. cultural places (museums, art galleries, ...)
4. leisure areas (sports equipment, restaurants, ...)
5. squares and open spaces

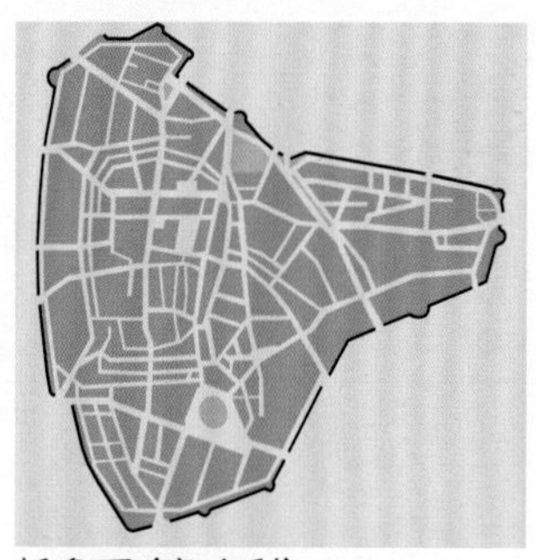

帕多瓦（意大利）
Padova (Italy)

历史博物馆
1. historical monuments

宗教中心
2. place of worship

文化中心
3. cultural places

休闲区
4. leisure areas

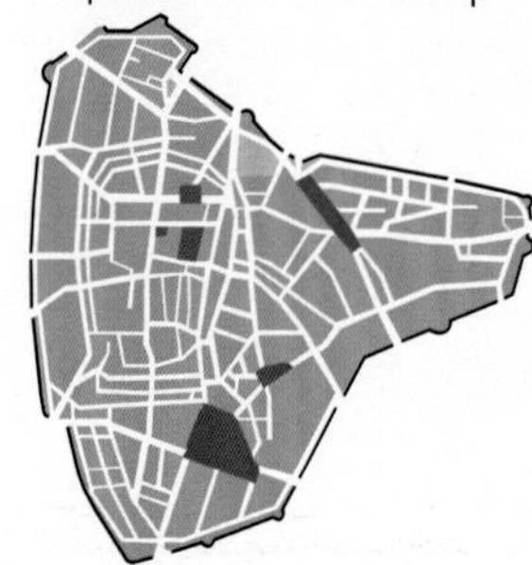

开放空间
5. open spaces

虚与实

从文化意义来说，实有积极的含义（富足），而虚有消极含义（缺失）。

而在城市概念中，“虚”有着积极的含义。
城市中虚与实的空间标志着不同的片区，而且有助于城市各片区的互动和依赖。
实代表城市中已经开建的部分，而虚表示还未开建的空间。

间隙可以是：

- 公共空间（街道、人行道、公路）
- 公园（绿地）
- 未城市化的区域（工业废弃区、待复原区）

FULL AND EMPTY

Culturally, the concept of full is linked to a positive sense (plenty), while the vacuum a negative sense (loss).

Within the urban concept, empty instead only takes a positive value.
Within the city, the relationship between the full and empty spaces identifies the neighborhoods and generates contacts and real dependencies between parts of the city.
The concept of full identifies the built-up parts of the city, while the concept of empty identifies spaces unbuilt, the gaps.

Gaps can be:

- public space (streets, sidewalks, roads)
- parks (green areas)
- not urbanized areas (brownfields, areas to recover, ...)

实心图
full

空心图
empty

双向两车道城市街道

urban street
with two lanes

城市街道＋人行道

urban street +
pedestrian routes

城市街道＋行人道和自行车道

urban street +
pedestrian and cycle
paths

通道

路径与虚实概念紧密相关，路径通道实际上是城市空间里的组成部分。

为了保障高质量的生活和城市生活的充分互动，路径设计必须以高效便利的方式与各名胜地相互连接。

路径通道主要有以下几种：

- 机动车道
- 人行道
- 自行车道
- 航道

不同类型的通道可以组合在一个线性方向通行，或者用隔离带隔离，或者混合在一起通行。

从更广泛的意义上来说，一部分通道成了绿色走廊，与城市绿地相连，在城市建筑系统中成为绿地网络的一部分。

不同类型的通道可以按以下因素分成不同的系统：

- 重要性
- 交通流量

PATHS

Trails have a close relationship with the concept of full and empty; paths in fact fit within urban spaces.
To ensure a high quality of life and a high degree of socialization in direct contact with the city, it is essential that the paths connect nodes and the points of interest between them in an efficient and attractive way.

We can classify paths as:

- roads for motor vehicles
- pedestrian paths
- cycle paths
- waterways

The different types of paths may occur together in a single axis or with a separate or mixed system.

In a broader interpretation, green corridors also are part of the paths that put the green areas of the city in communication with each other, generating a green space network within the system of the buildings.

The different types of routes can be classified in a hierarchical system based on:

- degree of importance
- traffic flow

通道网络可以按以下几个标准来划分：

- 根据空间指向
- 根据通行目的（通勤、购物、旅游，等等）
- 私人通道或公共通道
- 根据通道基建类型（公路、铁路、航空、航船）
- 根据运输介质

The networks of paths can be classified:

- according to the spatial reference
- according to the purpose (commuting, shopping, tourism ...)
- private or public
- according to the type of infrastructure (road, rail, air, waterway)
- according to the type of transport medium

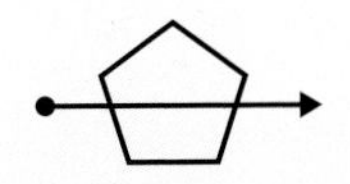

穿过
through

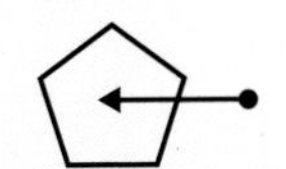

进入
input

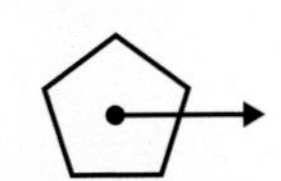

输出
output

内部流通
inside

3 未来的城市
现在的规划
城市设计（再设计）- 标准

3 the city of tomorrow
planning today
(re)designing cities - criteria

3 未来的城市

我们周围的空间和土地正在以一种失控的方式快速地发生改变。

在这样一个历史性时刻，我们有机会改变这种局面，让城市转型有明确的方向。

要做到这一点，我们必须找到改变的关键点，也是未来城市的支柱要素：

- 节约能源
- 保证生活质量
- 城市密度

上世纪城市人口呈爆炸式增长。
根据分析结果，到2025年，预计70%的世界人口将居住在城市；因此我们要将城市建设得更适宜人们居住。

建筑虽不会产生污染但建筑本身也不能变成能源。
我们在设计规划的同时应该用一种更合理的方式去思考。

3 THE CITY OF TOMORROW

The territory and the space around us are changing at breakneck speed and in an uncontrolled way.
In this historical moment, we have a great opportunity to change course and give a direction to this transformation.
To do so, we must focus on what we consider the key points for change, which are the pillars of the city of tomorrow:

- NO WASTE OF ENERGY
- PROTECTION OF THE QUALITY OF LIFE
- URBAN DENSIFICATION

Over the last century, the population living in cities has grown exponentially.
According to analysts, to the year of 2025 , 70% of the world population will live in cities;
therefore, we have to make our cities more livable in relation to the steady increase in the settled population.

The buildings will not pollute and will themselves be a source of clean energy.
We will have to rethink in a more ethical way how to act on our territory.

城市居住人口
population living in the city

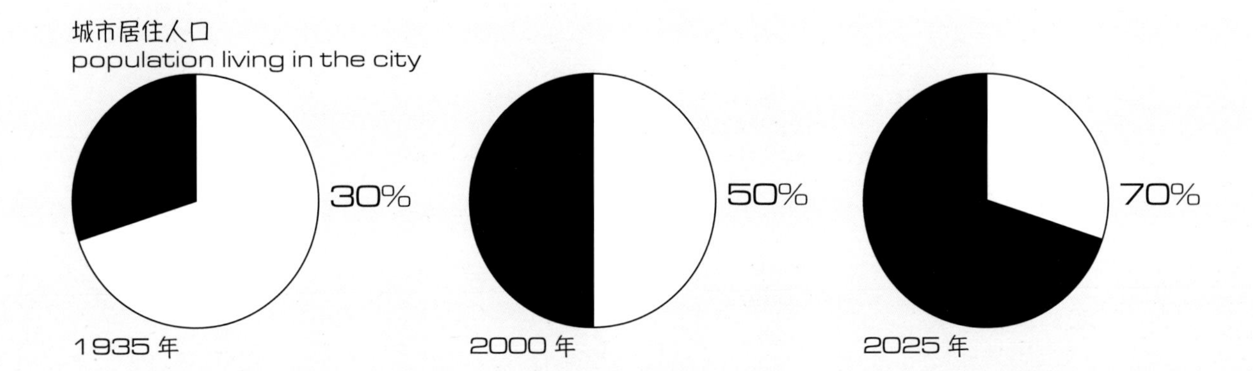

现在的规划

现在的规划设计是建设未来城市的巨大机会。
规划意味着相信未来，心怀远见。
规划设计没有统一的解决方案，每个城市像每幢建筑一样因人不同而不同，因地不同而不同。
城市化过程是一个长时间的过程，只有未来的人们才能看见它的结果。

合理的规划需考虑以下两方面：

- 能源与土地的关系
- 城市与用地的关系

能源与土地的关系

我们自身与能源和土地的关系是大家必须思考和践行一个重要问题。
能源和土地是宏观系统中的构成部分，是非常稀少和有价值的商品，所以我们必须节约使用。
它们是两个相对高度独立的系统，因此系统之间的互相干预会造成对各自系统的影响。上世纪我们按照低能源成本的理念开发土地。（城市以平铺的方式扩张）在发达国家，总能源的30%是用在建筑上的，而交通运输也消耗同样的能源。

为了合理利用能源，我们必须大幅减少以下用量：

- 人、货、服务的实际运送距离
- 用于建筑供暖和降温的能量消耗

只有这样才能降低全球能源的消耗量。

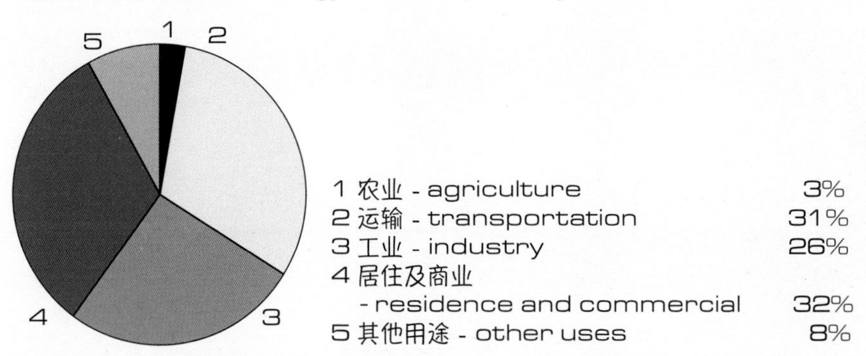

PLANNING TODAY

Planning is now a huge opportunity to configure the cities of our future.
Planning is to believe in the future and have visions.
There is no unique solution; every city (just as each building) is different, because every people and every area are different.
The processes of change in the urbanized territory unfortunately takes a long time; the results can only be seen by future generations.
For proper planning it is necessary first of all to take into account two aspects:

- the relationship between energy and territory
- the relationship between city and land

THE RELATIONSHIP BETWEEN ENERGY AND TERRITORY

Our relationship with the energy and the territory is one of the fundamental issues on which to reflect and act.
Energy and territory constitute a single macro-system and are rare and valuable goods: they must be used sparingly.
They are highly interdependent and, therefore, every intervention on one of the two systems has repercussions on the other.
Over the last century, we have developed our territory in accordance with the principle of low cost energy (horizontal urban sprawl).
In developed countries, about 30% of the total energy consumption is absorbed by the buildings and as much for transportation.
To promote a rational use of energy, we will need to drastically reduce:

- physical distances for the transportation of persons, goods and services
- the energy consumption of buildings for their heating and cooling

thus cutting a considerable slice of the global energy consumption.

城市和用地的关系

我们在土地上建起了大型购物商场，但是对这片土地却没有任何情感眷恋。
我们的将来有赖于我们生活的这片土地的将来，而这片土地的将来就与我们城市的未来直接相关。

很不幸，全球化的逻辑似乎是越来越关注城市及其周边地区，以及它们二者之间的关系。

这样一来，我们似乎与土地越来越疏远。

欧洲南部城市实施的规划结果看起来惨不忍睹，原因主要是：

- 历史的积重
- 缺少规划（杂乱）
- 缺少决策
- 缺少指引和目标
- 普遍追求经济利益
- 对未来缺乏信心

以上这些原因导致城市向一个错误的方向变化，使整个城市的理念和机制都成为失败。也许我们都忘记了我们的生活是多么美好。

126

THE RELATIONSHIP BETWEEN CITY AND LAND

We are creating disaffection towards our territory, which we are turning into a huge shopping mall.
Our future depends on the future of the area in which we live; the future of the area in which we live is linked to the future of cities.
Unfortunately, the logic of globalization seems increasingly worrying for cities and their surrounding regions, and the relationship between them.
In so doing, we are creating disaffection towards our territory.
The present results of city planning applied to cities in southern Europe remain disastrous for a number of reasons, such as:

- the weight of history
- the lack of planning (chaos)
- the lack decision-making
- the lack of guidelines and specific objectives
- the prevalence of economic interests
- the lack of confidence in the future

The combination of these factors has led to the evolution of the city in a wrong way, contributing to the failure of the institution and the concept of the city itself.
We have forgotten that in our cities it was and still often is beautiful to live.

城市不仅是由物质元素组成的，它也是人类社会的缩影，是有着丰富体验的染缸，是激发创意的温室，是人们相聚的地方。

城市和周边的土地决定了我们是什么，我们怎么生活。

21世纪面临的挑战就是重新规划我们现有的城市，从一个新的角度重新思考，试着弥补我们过去犯下的错误。
这些修正过程都必须向着特定的目标进行。

城市的规划设计和建筑都必须做到：

- 经得起时间考验
- 最大限度减少维护成本
- 减少浪费
- 逐步调整
- 让城市与土地、气候、历史、文化相适宜（目前来说，赤道和西伯利亚地区的新区规划不太可能有变）

The city is composed not only of physical elements,
but it is also an epitome of man, a melting pot of experiences,
an incubator of ideas and a meeting place.
The city and the territory surrounding it determine
our way of living and being.

The challenge of the twenty-first century will be to redesign
the existing cities, to rethink them in a new perspective,
trying to remedy the mistakes of the past.
This process must be modulated according to specific goals.

The output of city planning, urban design and construction
must:

- stand the test of time
- have maintenance costs reduced to a minimum
- allow us to avoid waste
- lead to a gradual readjustment of the city to its territory, climate, history, culture (at present the new districts of the cities Equatorial and Siberian are virtually identical!)

127

宏观城市
macro system city

=

宏观建筑
macro system building

城市设计（再设计）- 标准

城市再设计的原理和建筑设计的原理是一样的。

重新规划的未来城市须做到：

- 在能量供应方面可以自足（独立的原理）
- 不产生垃圾（道德伦理原则）
- 使居民享有幸福生活（质量原则）

追求这些目标将是新千年的挑战。

要实现以上目标需关注以下几个重要概念：

- 人口密度
- 灵活性和多目的地的混合
- 虚置空间
- 行人空间
- 公共交通优先
- 合理利用能源
- 让公共空间有趣而有激情
- 社交模式
- 消除贫民窟
- 社区个人双赢

(RE)DESIGNING CITIES - CRITERIA

The principles on which the redesign of cities is based are the same that underlie the design of buildings. The future city "re-designed" in the twenty-first century will have to:

- be energetically self-sufficient (principle of independence)
- not produce waste (ethical principle)
- allow a happy life for its inhabitants (principle of quality)

Pursuing these points will be the challenge of the new millennium.

To succeed in this aim it is necessary to focus on the following key concepts:

- densification
- flexibility and mix of destinations
- empty spaces
- pedestrian permeability
- priority to public transport
- rational use of energy
- make public space interesting and stimulating
- social model
- avoid ghettos
- win-win model community - private

人口稠密度

鼓励竖向发展。

将人类的未来寄托在空间迁移是不可能的，除非我们能找到一种新的低成本能源。

我们需要将人类更集中，以满足在有限的土地上人类的居住需求。

城市地区的蔓延扩张理念应该摒弃。

DENSIFICATION

Vertical growth must be encouraged.

Being impossible to base the future of humanity on physical movement (until a new source of energy at low cost is found), we need to concentrate the people in clusters of higher density, which are designed in such a way so as to meet the needs of people with limited displacement.

The concept of urban sprawl must be abandoned.

128

129

多个目的地相结合及灵活性

同一个地方或同一幢建筑能履行多个功能可以保证邻近地区的人流移动，而且能保证每个社区都能得到货物和服务，从而避免长时间出行，减少能源消耗。

MIX OF DESTINATIONS AND FLEXIBILITY

The inclusion of different functions in the same area
and in the buildings ensures movement within neighborhoods throughout the day and the procurement of goods and services within each district, thus avoiding the need to travel all the time and decreasing energy wastage.

130

131

开放空间

城市开放空间主要有以下几种：

- 铺设的公共空间（街道、人行道、公路）
- 公园（绿地）
- 未城市化片区（待开发片区）

这些都是市民们必定会体验的开放空间，是他们聚会、互动和休息的空间。

虚空的结构和有序的空间是城市的附加值所在，这些空间增加了城市的活力，给人们带来愉悦的感受。

我们应该实施城市空间再开发的良性规划，让城市空间成为新城市的动力激活器。

EMPTY SPACES

The empty spaces in the city are:

- paved public spaces (streets, sidewalks, roads)
- parks (green areas)
- spaces in non-urbanized areas (areas to be recovered)

Those are all empty spaces that must be experienced by the citizens and take the function of meeting spaces, of places for rest and interchange. The hollow-structured and organized spaces add value to the city, increasing the livability and a positive population's perception.

A virtuous process of re-evaluation and redevelopment of urban voids must be undertaken, to make them the vitality activators of the new city.

132

133

行人分布

我们应该鼓励行人分布。
市民们应该可以不受空间限制，在城市里自由行走。

行人必须能够通过行人活动区域，如行人通道、自行车通道等，自由行走到城市的每个地方，并且不会受机动车交通的影响。

行人分布主要用自行车出行和步行，可以推进“绿色出行”。

PEDESTRIAN PERMEABILITY

Pedestrian permeability should be encouraged.
Citizens must be able to move within walking distance in their city without spatial limitations.

People must be free to walk to every area of the city through the creation of dedicated pedestrian areas, cycling and walking paths separated without incurring the dangers of vehicular traffic.

The wise use of pedestrian permeability promotes "green mobility" through the use of cycling and walking.

134

135

公共交通优先

现在的城市距离不再用公路里程数而是用花费的时间来衡量，因此，有必要建立一套高效的交通运输系统。

鼓励大家用公共交通出行，这样可以减少通行时间和汽车尾气污染。

PRIORITY FOR PUBLIC TRANSPORT

It is necessary to ensure a high transport efficiency since distances in the city are no longer measured by the metric system but based on time.

By encouraging public transport on a large scale, it is possible to reduce journey times and therefore pollution.

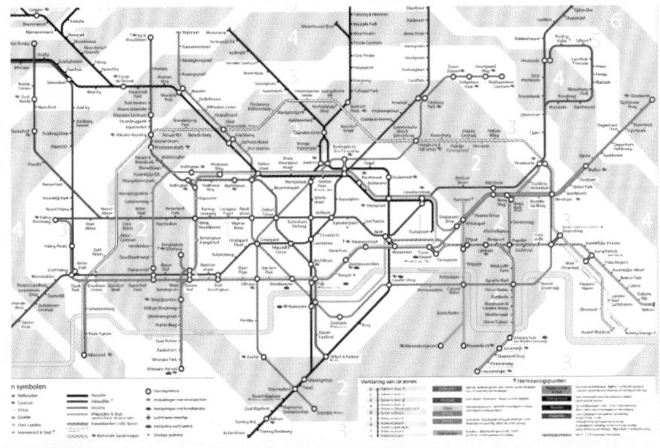

136

137

合理利用能源

RATIONAL USE OF ENERGY

目前的能源结构在短时间内注定会崩塌，但是新的可替代能源还未找到。

The current energy model is destined to collapse in a short time and has still not found a suitable alternative.

核燃料的生产开发可以解决未来十年的能源问题，但是到2060年才可能实现。

Exploitation of nuclear fusion at an industrial scale could solve the energy problem that we will have to face in the next decades, but it will not be ready before 2060.

现今建筑的取暖降温系统浪费大量能源，对我们现在的生活，甚至将来的生活都会造成污染、缺水、全球变暖等不良影响。

The current system for heating and cooling of buildings, based on the waste of energy, creates negative consequences for our present and future life: pollution, water scarcity, global warming.

未来的人们必须利用各种技术来降低能源消耗，不管是从微观方面还是宏观方面都必须自觉朝这个目标努力。

Those who work, move, or construct in the future must be driven to use the available technology to reduce energy consumption, with the aim to achieve autonomy at the micro and the macro scale.

138

139

让公共空间富有创意，充满激情

我们必须让市民们，尤其是年轻人，喜欢去公共空间，从而培养起他们对那里的文化和历史的热爱。

这样才能让公共空间焕发活力，真正为市民所用所享。

MAKE THE PUBLIC SPACE INTERESTING AND INSPIRING

We must convey to the citizens
and especially to the young generation
a passion for public places
(and therefore for the culture,
for the specificity of the place
and its origins)
and the desire to go there.

In this way, it is possible to revitalize
public spaces and return them
to the citizens.

140

141

社交模式

一个人需要跟别人打交道，分享别人的经验才能成长，因此人们需要社会交往。

社会交往的场景主要有两种：虚拟模式和现实模式。

SOCIAL MODEL

To grow, man needs to deal with others and to share their experiences: people need to socialize.
This need manifests itself in two possible scenarios: the virtual model or the real model.

142

143

虚拟模式

在虚拟社交模式中，人们显得越来越“形单影只”，因为人们只是通过虚拟的方式互相联系。
（“黑客帝国”与虚拟现实）

Virtual model

In the virtual model people are more and more "physically alone", only virtually connected to each other.
("Matrix" and virtual reality)

144

145

现实模式

按照现实模式的场景，社交应该是真实可触的：情感心智完备的人们在一个真实的地点相聚，进行真实直接的情感交流。

我们相信大家还是更愿意参与到实实在在的人与人之间的互动沟通中，所以，我们也需要实实在在的空间相聚交流。

为了促成这种“现实”的模式，我们需要设计一些有意思的、方便人们开展各种活动的地方，吸引人们花时间在这里和他人交流。

Real model

According to the real model, socialization is based on straightforward feelings and emotions: the people, each individual in his physical, mental and emotional completeness, come together in physical locations.

We believe that every individual is still in need of the three-dimensional body to interface with each other, so we need physical places to meet.

In order to push towards the creation of "real" models, we need to create physical places of interest, representing a stimulating and attractive frame for the activities of man, inviting citizens to spend time with others.

消除贫民窟

AVOID GHETTOS

世界上很多地方不断有新的“封闭片区”出现，成为现代贫民窟。

Everywhere around the world
new "closed quarters"
are being created:
the contemporary's ghettos.

因此，贫民窟必须消除，因为它会自动让社会分化和隔离，引发分裂主义，破坏西方民主社会的根基。

The ghettos, therefore,
must be avoided because
they automatically generate diversity
and social segregation phenomena
triggering separatism and breakdown,
undermining the foundations of the
occidental democratic society.

146

不管富人的地产周围是否竖起保护隔离带，但现代贫民窟依然在破坏代表城市财富和灵魂的不同团体和不同文化之间的交流互动。

Whether spontaneous
(the favelas, ghettos of the poor)
or areas fenced to protect the assets
of the rich, these contemporary
ghettos are destroying
social interaction and interchange
between different groups
and cultures, which represents
the wealth and the soul of the city.

147

社区和个人双赢模式

WIN-WIN MODEL COMMUNITY - PRIVATE

城市设计的负责人必须心怀大众，毫无私心。

The director of operations of the city must remain public: this task cannot be delegated to private initiative.

整个规划修整也应该由大众参与开发。
个人的利益应该符合大众整体计划的界限。
在这样的框架下，最重要的一点是要有资本和金融工具的支持。

The general plan of intervention must always be developed by the public. The private fits with its initiatives within the public general plan, which sets the limits and methods of intervention. In this framework, the fundamental role is played by the availability of capital and financial instruments.

148

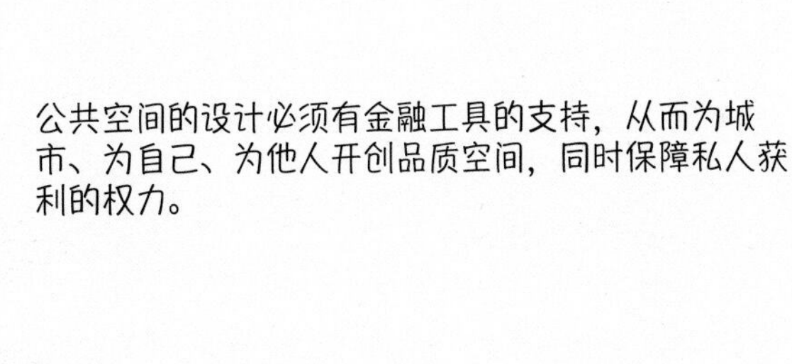

公共空间的设计必须有金融工具的支持，从而为城市、为自己、为他人开创品质空间，同时保障私人获利的权力。

Those working in the public must be provided with the financial instruments in order to achieve the quality spaces for the city, for himself and for the people, and at the same time guarantee the right to private gain.

149

公共_私人组合的方案很受欢迎，但是管理控制，包括执行权应交由公共部门来开展，其行动及后续的效应需基于民众的立场

Mixed public-private initiatives are welcome, but control, including execution, must be exercised by the public sector, which remains responsible for its actions and their consequences in the eyes of the citizens.

4 案例
迪拜
圣加伦
哥本哈根

4 examples
Dubai
St. Gallen
Copenhagen

迪拜

迪拜（阿拉伯联合酋长国）是城市发展策略错误的典型案例，而且城市设计与土地之间关系不协调。它是一个只靠丰富石油资源而盲目发展的大都市。

DUBAI

A city epitomizing a wrong
development strategy
and an inappropriate relationship
with the territory is Dubai
(United Arab Emirates):
a huge metropolis
whose indiscriminate development
has been based exclusively
on the abundant availability of oil
in its underground.

150

151

按照低能源消耗的准则来考虑的话，迪拜的城市设计标准落后了。一旦酋长国的石油和天燃气断供，迪拜将像恐龙一样注定被埋葬。
耗费大量能源来淡化咸水，灌溉沙漠中的花床和高尔夫球场，保持室温凉爽而不用任何建筑隔热材料，这种模式不可能永久持续。

石油换来的美元和金融是这个城市唯一的给养，与这片土地没有任何灵魂相系。

Conceived on the availability
of low-cost energy, the city has been
designed on the basis of old canons:
Dubai is a dinosaur destined to run
aground as soon as the extraction
of oil and gas in the emirate will stop.
The cost of energy to desalinate water
that irrigates the flower beds
and golf courses created in the desert
and to cool down buildings w
ith almost no insulation
might not be sustainable forever.

The finance and petrodollars
are the only raison d'être
of a city that does not have a soul
linked to the territory.

152

153

与领地没有任何关联——只有被需求驱动的流转

迪拜这座城市与周围的环境没有任何关联，没有社交联系组织结构，城市构成单位各个孤立，相距甚远，城市的运营和维护需要很大花费。

城市发展与海洋毫无关系。

海洋与居民区的联系也仅限于几座奢侈的复式居民楼。

NO CONNECTION WITH THE TERRITORY - ONLY DRIVEN MOBILITY

Dubai is a city with no connections to the surrounding context;
it is a city without a social connective tissue, consisting of isolated incidents distant from each other that require a strong commitment to their operation and maintenance.
The city developed in the absence of any relationship with the sea;
the sea gets in contact with the residential areas only in correspondence with a few luxury residential complexes.

154

155

在迪拜出行只能靠汽车，宽阔的林荫道只为机动车通行设计。
公路系统无法让行人通行。

Dubai has been conceived for the car.
The city is traveled only by car;
the wide boulevards are only designed for motorized traffic.
The road network does not allow any pedestrian permeability.

156

157

没有灵魂的大型商业大楼

LARGE COMMERCIAL MALLS - THE SOUL IS ABSENT

城市的聚集地点就是购物中心了：大片建好的土地毫无个性可言，人们对这样的地方也没有特别的眷恋，因为这里的建筑跟欧洲和美国城市的建造风格一模一样。

The meeting places of the city correspond with shopping centers: large covered areas without their own identity, which combine to create disaffection to the site. As they are built here, they could be made in the same way in any European or American city.

城市里一有风吹草动人们就争相逃离：为数不多的迪拜居民飞往国际化西方城市，大部分外国劳工飞往他们原来的国家。

Any even slight crisis would have the effect of exodus: the few citizens of Dubai would fly to the international Western cities; the vast majority of the population, consisting of a foreign labor force, would fly back to their land of origin.

实际上，没有人在迪拜种下"情感的根"，因为这片土地并不能让人激发出这种与城市之间的情感联系。

In fact, no one has planted "emotional roots" in Dubai due to the inability of the area to encourage the establishment of links with the city.

158

159

贫民窟的天堂

这座城市的特点是每个地方都是一样的。干道沿线是酒店和奢华的写字楼，其后面是数英里的民工低矮住房区。

一些特别的购物中心和主要休闲区都是为富人而建的。

海水进入城市的地方就是迪拜河，是主要的旅游景区，老城的中心区。那里有一些服务游客的店铺和船舶。

都市发展和大量资本的进入造就了一个富人贫民窟。

另一个和这个体系惊人相似的地方是棕榈岛。这是阿拉伯海湾里凭空建起的几条狭长的陆地，上面全是高档酒店住房。去岛上任何地方，如商店、聚会点、酒吧、饭店都很远，必须开车前往，无法靠步行到达。

THE PARADISE OF THE GHETTOS

The city is characterized by homogenous areas corresponding to destination of use: the main roads are lined with hotels and luxury offices, behind which stretch for miles expanses of low housing units for migrant workers.

Some specialized areas contain shopping centers and the main entertainment venues of the city, always aimed at the rich.

Around the tongue of salt water lapping at the city, the Dubai Creek, are the main tourist attractions, the ancient core, the boats and the shops for tourists.
The urban development and the wide availability of capital led to the creation of ghettos for the rich.

A striking example of this system is the Palm Islands, strips of land created out of nothing in the Arabian Gulf where luxury accommodations were built.
From the houses on Palm Islands, people move only in the car: the distances to any location, to a store or meeting place or to a bar or restaurant, are too large to be covered on foot.

160

161

圣加伦

瑞士圣加伦中心的城市广场是一个非常成功的城市再设计案例。

该项目的基本出发点是想通过公共空间的再设计开辟出一个建筑与建筑之间相联的连接元素：公共空间的设计打破了传统的设计方式，用艺术的，又有创意的视角打造出一个单色调的动感设计。

整个公共空间铺设了一块红色的橡胶地毯，连接着周边的建筑。

地毯下覆盖了各种街道设置的家具，如桌子、长凳。这个广场成了人们可以常去的户外房间。

ST. GALLEN

City lounge is a successful urban redevelopment project conducted in the city center of St. Gallen (Switzerland).

The basis of the project is the desire to create a connective element between the buildings through the redevelopment of public space: public space has been redesigned in an unconventional way through an artistic and creative vision, based on the vibration of a single color.

The public area is entirely covered with a rubber flooring of red color that has the function of a connecting element between the buildings.

With the red carpet that stretches over all public spaces in the neighborhood, all the elements of street furniture such as tables and benches are also integrated. The plazas and squares are transformed into large outdoor rooms accessible to the people.

162

163

164

165

166

更重要的是，这种橡胶地毯很软，不容易让人受伤，可以放心让儿童尽情玩耍。

从这个项目我们可以看到，一些简单的改变就能恢复城市的活力，提升公共空间的品质，吸引人们游走，成为令人们愉悦的社交地点。

空间里横向竖向的标志都被简单明快的线条所取代，这就是创意带来的成功设计。

这个项目就是一个公共-私人双赢的完美实例。这个项目是由一个金融机构发起的，由公众管理控制的设计项目。

Important is the recreational aspect
generated from the pavement
by the softness and anti-trauma effect
of the red rubber and by the inclusion
of play opportunities for children.

In this project, we can see how
with a few simple measures
it is possible to restore vitality
and enhance the public space,
creating flows of people throughout
the day and making it a pleasant space
for the socialization of people.

The horizontal and vertical signage
has been replaced by graphic signs
clear and clean:
creativity has triumphed.

The project is a perfect example
of a win-win public-private partnership,
in which the intervention was directed
and controlled by the public
and sponsored by a financial institution.

哥本哈根

北欧自古以来就有高质量的城市绿化空间。在城市中心很容易看见公园，配有户外家具的草地，市民可以日常使用。

丹麦哥本哈根的Superkilen完美展示出：全面的城市规划能做到利用城市未开发空间并带动整个周边地区的活力。

这项设计是在一个多民族社区方圆一英里的地方展开的，用一些拼色和造型将那里变成一个公园和一个广场。

这个项目被认为是代表居民多元化的各国物品的一个收集工作，收集的物品直接运用到装饰环节。

COPENHAGEN

Northern Europe has a great tradition of high quality urban green spaces. In the city center it is not rare to find parks, grasslands with street furniture, made to be used daily by citizens.

Superkilen in Copenhagen (Denmark) is a perfect example of how careful urban planning can reuse an empty disused urban place and revitalize an entire neighborhood.

The action unfolds along a one mile strip in one of the most multi-ethnic districts of the city, transforming a park and a square into a patchwork of colors and shapes.

The project is conceived as a sort of collection of objects from around the world representing in different ways the multiplicity of local residents, who were directly involved in the selection of the new furnishing.

167

168

169

170

171

行人通道，自行车通道都与当地的交通网络相连，市场和户外休闲片区这些城市元素都画上了跳跃的颜色，整个设计同时也反映了这个区域文化多元的特点。

这种审美艺术手法注重展现一个人为环境，而不是让它假装成一个自然环境。

最有意思的是，原来整个社区像城市迷失的一角，而现在人们利用公共艺术来增进相互之间的交流互动，整个社区积极参与。

Pedestrian and cycle paths, connect to the local transport network, an area for the market and outdoor recreation areas are the elements with which the project has been able to play through the use of lively colors and the basic idea of reflecting the new multicultural image of the area.

The intervention strikes one because of the aesthetic approach that focuses on a decidedly artificial environment instead of pretending to be natural. Some of the most interesting aspects are the community involvement and the use of public art as a means of social inclusion to promote exchange and encounters between people in what was once only a lost urban space.

城市
CITY

建筑
BUILDING

产品
OBJECT

坎图-历史与现状

坎图是意大利北部的一个城市。这座城市因其精湛的手工艺以及建立意大利最早的家具和室内装修艺术学院而闻名。

坎图与米兰相邻，城市生活也深受米兰影响。城市的一大难题是交通混乱，道路严重受阻，不过这也是全球大都市普遍面临的一个问题。

坎图在上世纪80年代有着辉煌的手工业和工业，现在坎图仍以此为骄傲。但坎图还是并入到了布里安扎地区，而布里安扎是米兰的一部分。米兰是最吸引有活力、有梦想、有希望的年轻人的城市。

CANTU' - STORY AND SITUATION

Cantù is a city in northern Italy, known for its strong handicraft vocation and for the first art school for the furnishing of the nation.
Cantu's life is strongly influenced by its proximity to the city of Milan.
One of the major inconveniences for the city is the chaotic traffic conditions that strongly hinder public and private roads, as is true for all cities in the sphere of influence of a metropolis.
In the 80's, Cantu, still proud of its craftsmanship and industrial capacity, was incorporated into the "magma" of the so called Brianza region.
The Brianza region has now become part of the territory of the unformed greater Milan, attracting the energies, dreams and hopes of the younger and more dynamic part of the northern Italian population.

172

基本建设——公路网络图

虽然城市地位有所提升，城市本身又有一定的历史渊源而且极有可能转型为颇具吸引力的城市休闲中心，但城市的中心仍未得到开发，那里仍是渺无人烟，破败不堪的不毛之地。

造成现在这种境况主要是因为毫无约束、无孔不入的摩托车交通。
城市的公路网络极混乱，毫无脉络系统可言。

BASE - THE ROAD NETWORK

Despite the fact that it has a central historic area in an elevated position, potentially convertible into a real "city lounge", and because of the widespread infiltration of an intrusive and unchecked motor traffic, the centre of Cantù has not developed and has become a no man's land, unattractive and characterized by poor maintenance and the decline of its buildings and structures.
The road network of the city is permeated by chaos with no recognizable hierarchy.

173

环形路线——双环形

重新规划这座城市的理念包括开建两个封闭环形：内部环与原来的老城中心环有部分重叠，外部环成为20世纪建筑群区域的一个界线。

外部环的枢纽点有通往其他省的运输通道，通道沿线散布着一些公共停车场。通往中心区的路线也被搭建起来，但是入口有限制。

非常住居民从外部环进入中心区只能靠公共交通或是步行。
若与外部环相连则需建一小段隧道。

CIRCULATION - THE TWO RINGS

The concept of reorganization prepared for the city
by deltaZERO includes the creation of two complete rings: an
internal one, partially adherent to the limit o
f the ancient original core, and an external one,
to limit the edification of the 20th century.
The concept included the creation of public car parks
by the connections between the external ring
and the major axes to the center and the introduction
of privileged axes with limited access to the city core.
For non-residents the access from the external ring
to the city center can only take place by public transport
or on foot.
For the completion of the external ring,
the construction of a short tunnel section was required.

174

识别特征——界限和门户

为了更加突出市镇的风格，重拾以往的特色标识，外部环成了都市扩张的界线，环与市镇之间形成了一个带有走廊的土墙防御体。

设计师建议在外部环与进入城市通道的交叉点上可以搭建一些各有特色的主题门户。

用门户来展示各种不同的且享誉千百年的特色手工艺，也展示城里居民们从事的手工艺特点。

IDENTITY - THE LIMIT AND THE DOORS

To strengthen the idea of a city with a soul, and to give back to Cantù its lost identity, it was proposed to use the external ring as the limit of the dense urban expansion, creating an elevated rampart with walkways between ring and city. At the entry points to the city from the external ring the creation of doors with different characteristics (from the portal to the mere artistic installation) was proposed, which addressed the different artisan specificity and skills that gave fame to the city and for centuries characterized the work and the life of its inhabitants.

175

新生活——展示中心

我们建议将城区再细分成与主要公路网有联结的片区。

每个社区都应有一个提供城市各项服务和设施的中心区，如邮局、银行、食品店、酒吧、饭店及休闲设施，等等。

社区的划分范围以步行至城市中心的时间为准：从城市的每个点到城市中心的步行时间在15分钟内为宜。
因此，我们建议设计步行路线，保证行人流动性。

NEW LIFE - DISTRICT CENTERS

It was proposed to divide the urban area into districts, each one lapped by roads of the main network.
Each district shall be characterized by a center with all the urban services and facilities, from the post office to the bank teller.
In addition, in the district center a food shop must operate and bars, restaurants and leisure facilities must be created.
The creation of districts is based on the pedestrian travel time to the city: from every point in the city, a citizen should be able to walk to a district center in less than 15 minutes.
Consequently, the creation of a network of routes was proposed to ensure pedestrian permeability.

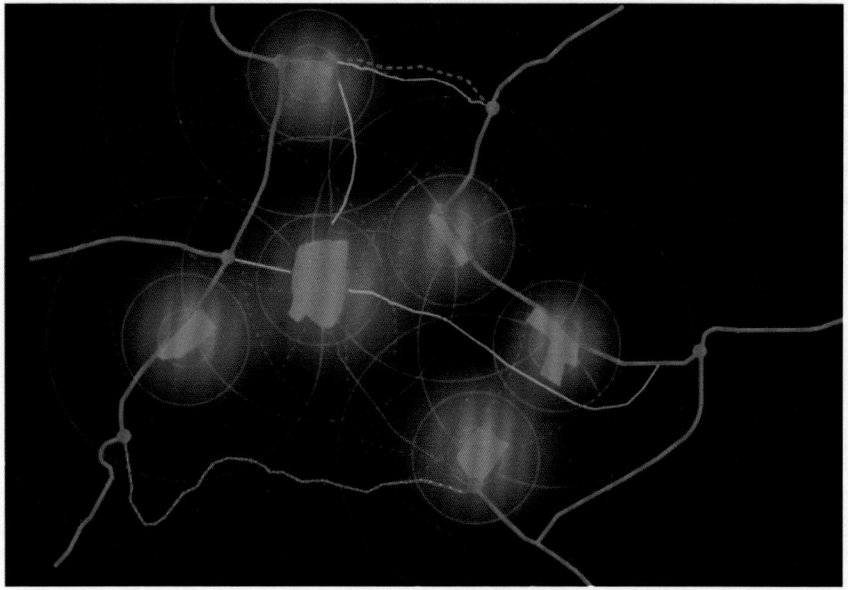

176

城市——模型

城市的设计模型都有一个基本理念，那就是让市民们能在一个轻松而有活力的环境里步行。
以一段市民愿意步行的距离算起（受文化、地形学等因素影响）建立起一个网格，网格是否规整依地形特点而定。
网格就嵌入在“交通环“里。

公路与交通环交叉的地方应具有以下条件：

- 设有停车场
- 与核心区、周边核心区联络通畅
- 可作为一个港口，特别入口，并担负一些发布交通环情况、城市居民特点等消息的任务

还城市活动和休息空间，让市民成为他们的空间的主人，对他们的空间怀有情感，这是当今努力设计好城市的首要任务。

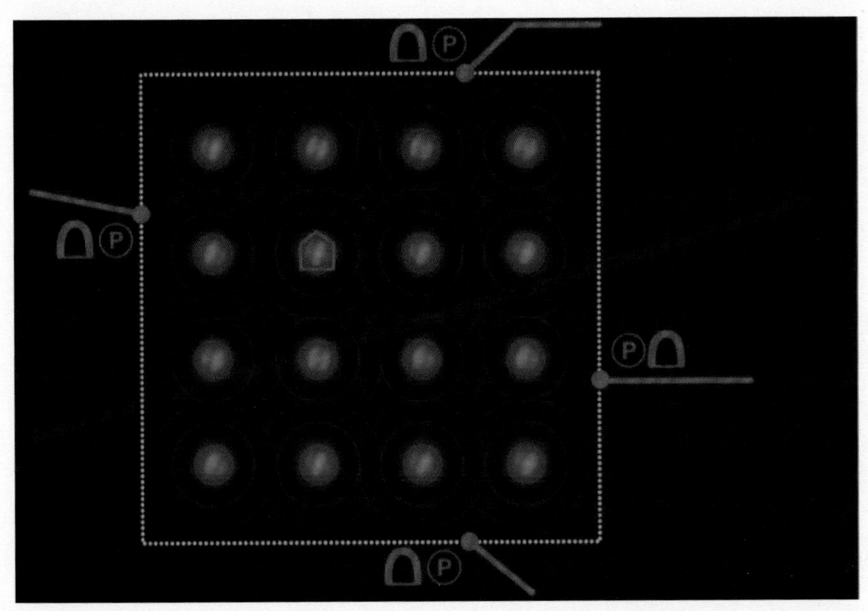

177

CITY - THE MODEL

The model that is the basis of the proposals
for the city of Cantù is based on the belief
that it is necessary to give the citizen the opportunity
to travel on foot, in a stimulating and relaxed environment.
Starting with the distance that the citizen
is willing to travel by foot (distance depending on several
factors, from the culture to the site's topography)
a grid was established, regular or irregular according
to the characteristics of the territory.
This grid is inserted within a circulation loop.

Each point of intersection between the circulation loop and
the axes of penetration must

- be equipped with parking
- have an efficient connection to the core and the district centers
- be considered as a door, a privileged place of access, and be appointed to provide information on the characteristics of the city and its inhabitants

Giving space for movement and rest within the urban area,
making the citizen master again of the spaces,
opportunities and emotions
that the urban environment offers is now the priority
for those who work for the good of cities.

城市
CITY

建筑
BUILDING

产品
OBJECT

DeltaZERO大楼，卢加诺

卢加诺的DeltaZERO大楼是与他齐名的建筑公司的总部。这座建筑的落成向人们展示了利用现有的技术，即使在需要供暖制冷的气候条件下（冬季气温低于零摄氏度，夏季气温高于30摄氏度，湿度超过80%）也可以保证无碳排放。

大楼的设计师有意将周边的城市环境融入大楼的生活，这也是这家建筑公司的设计理念（依托技术，灵活机动，具备职业道德，透明清楚）。而这样的理念化为现实就是这幢全玻璃幕墙，零能源消耗的建筑大楼。

DeltaZERO大楼是建筑师，工程师、以及能源、外墙、电力等各行业技术专家共同合作努力的成果。

178

THE DeltaZERO BUILDING, LUGANO (CH)

The DeltaZERO building in Lugano, site of the eponymous architectural firm and residence of its founders, is a prototype designed and built in order to demonstrate that, with the currently available technologies, it is possible to make carbon neutral buildings even with climatic conditions that require both heating and cooling (temperatures below zero in winter and above 30° C in the summer, and with humidity over 80%).

Given the intention of the designers to integrate the surrounding urban landscape into the life of the building and the philosophy of the architectural firm for which the building was designed (based on technology, flexibility, ethics and transparency), the goal was in fact, to achieve a fully glazed building and zero energy consumption.

DeltaZERO is the result of close cooperation between architects, engineers and technical specialists from many different industries, from energy to facade technology, from plant engineering to electrotechnical systems.

为了达到既定目标 DeltaZERO大楼（全玻璃幕墙，无遮阳保护）具有以下特点：

- 极高的绝缘隔热系数（Ug = 0.5 / g = 0.23）
- 地热系统（6x4管井，深130m，> 2.5km长的管道）还有一个双用热泵
- 一套带可供热和地面收集器的低压充气系统
- 用半活化状态的混凝土板来稳定室内空气温度
- 朝南墙面全部铺贴太阳能板，为大楼供应热水
- 屋顶集中了全部的光电伏系统
- LED技术支持的公共区域照明系统（600 W，7层楼及车库的照明）

大楼供热制冷耗能估计会达到27000 kwh 每年(18千瓦时/每年每平方米)，而这些能耗都可利用太阳能解决。

2009年DeltaZERO大楼启用，既可以办公也可以居住。

DeltaZERO大楼向大家证明了传统建筑（30% 的玻璃墙及外部遮阳箱）的能耗成本完全可以通过现有的技术解决。

179

To achieve the settled goals, the DeltaZERO building (fully glazed and without external sun protection) was equipped with:

- a highly insulating “shell” (facades and roof) with unique insulation and thermal characteristics (Ug = 0.5 / g = 0.23)
- a geothermal system (6x 4 pipe-wells, depth 130 meters, > 2.5 kilometers of pipes) and a reversible heat pump
- a low-pressure air-inflating system with heat recovery and ground collector
- semi-activation of the concrete slabs to stabilize the indoor climate
- solar panels, fully integrated in the south facade, for the production of hot water
- a photovoltaic power supply unit, fully integrated in the roof
- a lighting concept for common areas based on LED technology (with only 600 W, seven floors of stairs and the garage are illuminated)

The result of the technological effort is an energy consumption for heating and cooling of approximately 27000 kwh per year (only 18 kwh/year per square meter). The entire energy consumption of the building is covered by the sun.

The DeltaZERO building has operated as an office and residence since 2009.

DeltaZERO is proof of the fact that the coverage of total energy costs for traditional buildings (with glass surfaces comprising 30% of the total and with external sun protection casing) using current technology is easily attainable.

设计项目理念

DeltaZERO 大楼是一个全部采用玻璃幕墙的建筑，三面透明设计，可以欣赏湖景和城市全景，同时让人感觉跟随城市一起博动。
这是一幢零度耗能的建筑，是开启未来城市设计的第一步。
它的名字DeltaZERO，“零度”，意思是建筑消耗多少能源就能产出多少能源。

180 建筑外墙
181 关联

PROJECT IDEA

DeltaZERO is a fully glazed building, designed transparent on three sides to enjoy the view of the lake and the city and at the same time feel as a part of the pulsating city.
It is a zero energy building designed as a first step towards what could be the city of the future.
The name DeltaZERO, “differential zero”, means that the building produces as much energy as it consumes.

180 - building skin

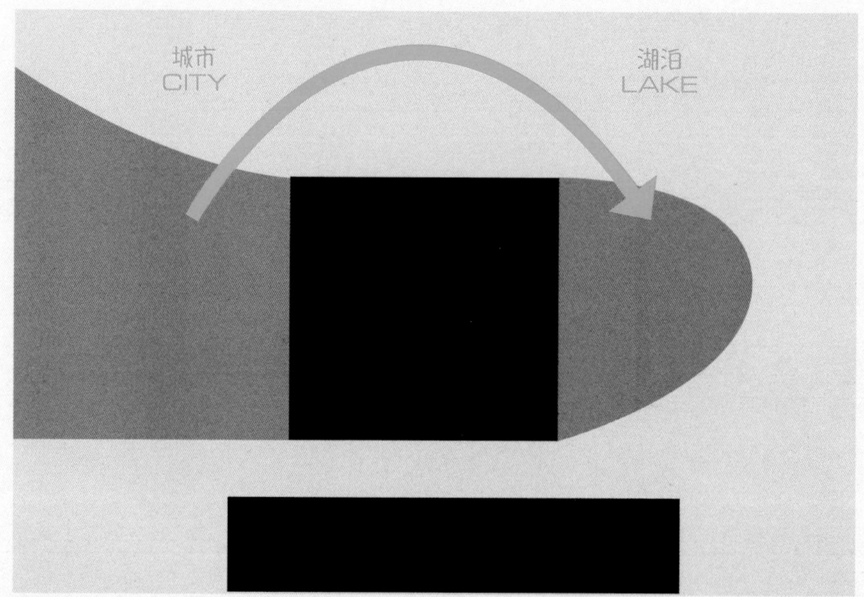

181 - connections

耗费能源 = 产出能源

DeltaZERO大楼的幕墙和屋顶都是玻璃材质，经过设计制作后既能满足建筑里的人们的舒适需求，同时又不会消耗能源，不释放有害物。

182 制冷制暖 - 27'000千瓦时 / 年
183 太阳能+ 27'000千瓦时 / 年

ENERGY CONSUMED = ENERGY PRODUCED

DeltaZERO is a building with facades and roof in glass and manufactured in such a way as to ensure total coverage of the energy required for the comfort of the building inhabitants and at the same time zero harmful emissions.

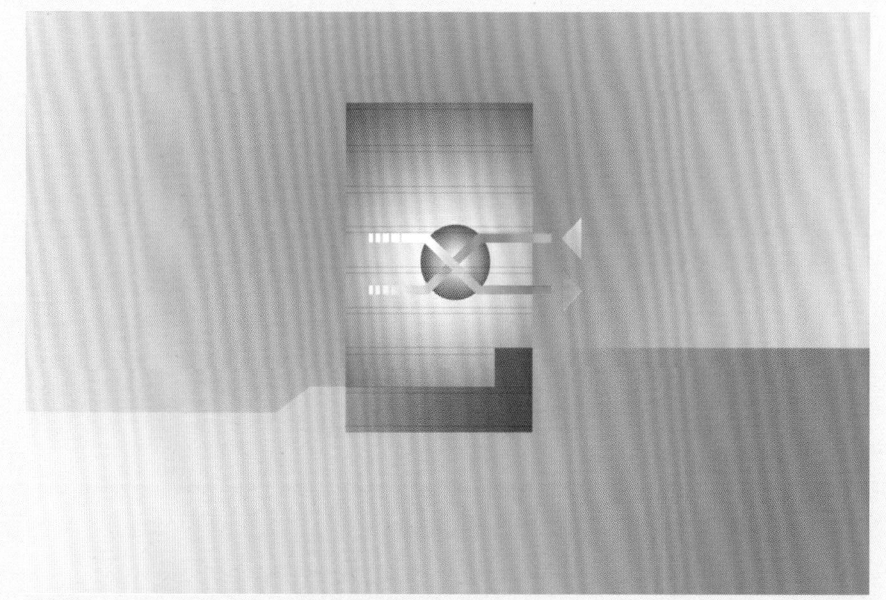

182 - heat & cool : - 27000 kWh/year

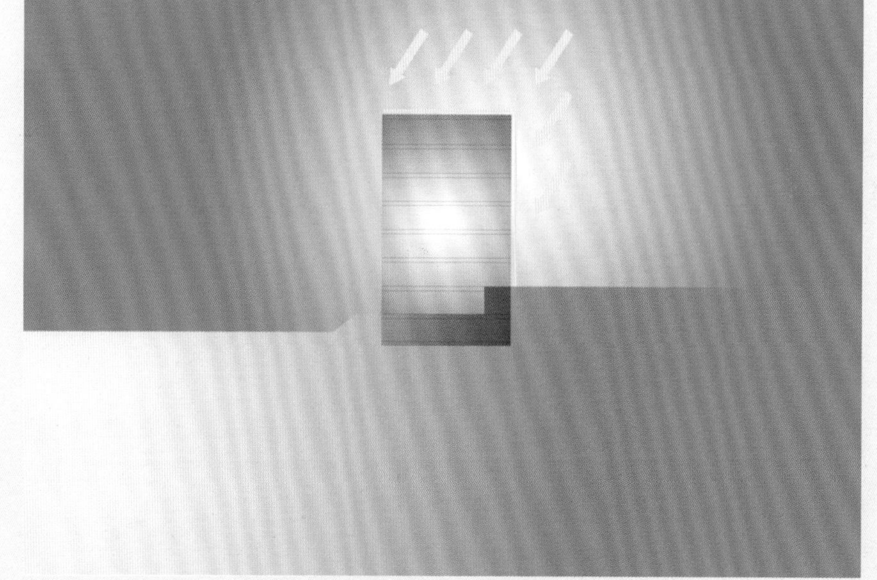

183 - solar energy production : + 27000 kWh/year

空间 & 灵活性

现代生活方式经常涉及空间利用。
DeltaZERO可以灵活运用建筑里的各个受力点。根据用户的需求，内部空间既可以用作办公室也可用作公寓。即使大楼峻工了，你也可以快速简便地改变内部空间及其分布。
为了方便大楼的技术系统及时更新，大楼里的供电供水及自动化系统都安装在活地板中。

184 存储间
185 办公室

SPACE & FLEXIBILITY

The contemporary living style requires constantly evolving spaces.
DeltaZERO makes flexibility the point of strength of the project.
The inside spaces can be used as offices or apartments depending on the immediate user's need.
Spaces and internal distribution can be modified in a fast, simple and inexpensive way even when the building is finished.
To equip the building with the newest technologies and to guarantee the possibility of making changes easily, the water, power supply and building automation are located in the raised floor.

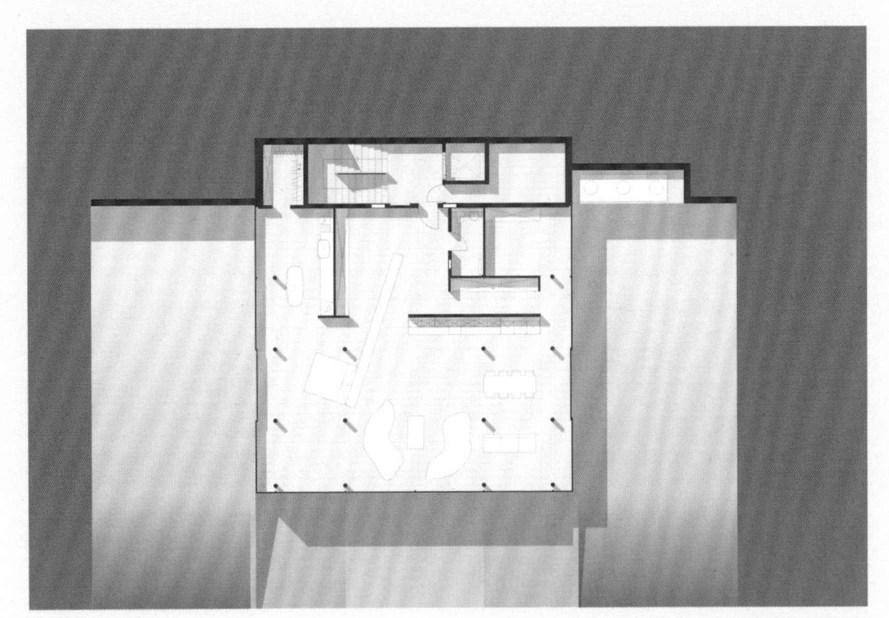

184 - loft

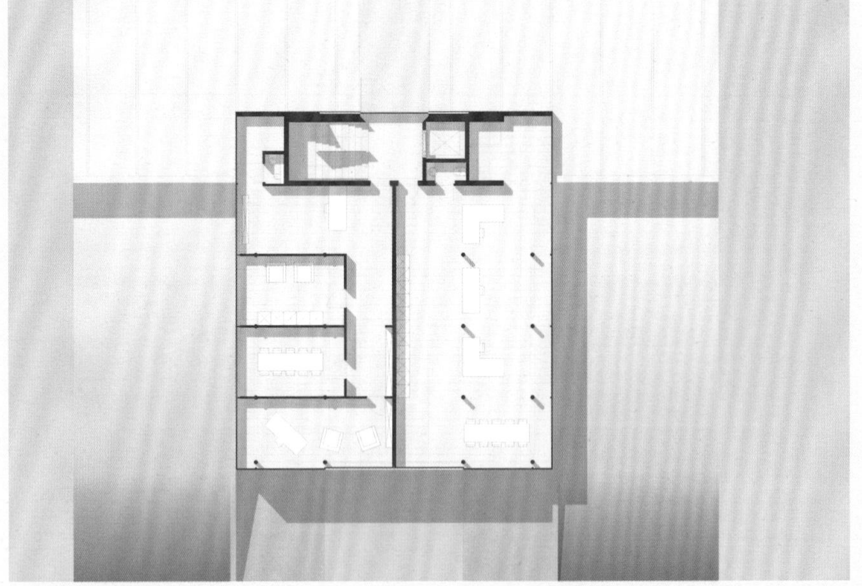

185 - office

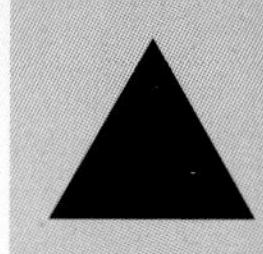

186 公寓
187 公寓

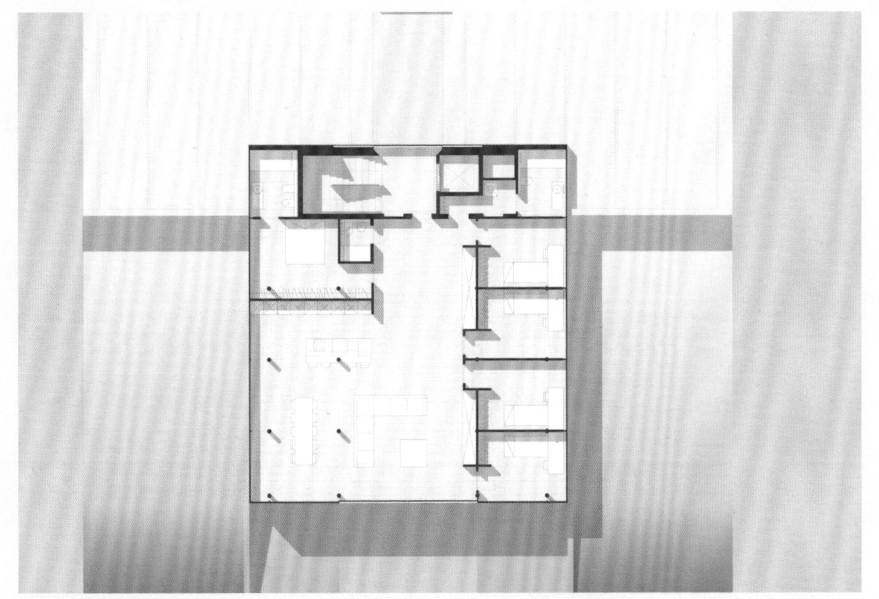

186 - apartment

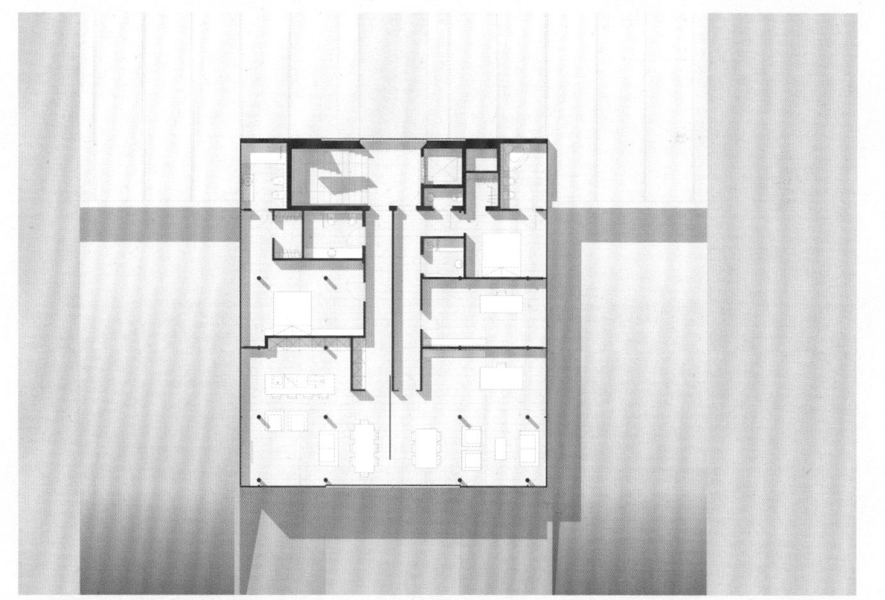

187 - apartments

主要结构——内部自由

有序的支柱结构可以让人们自由地改变分隔空间。
用加固水泥建造的楼梯可以起到加固整个结构的作用，缓冲水平方向的受力（风、地震）。

MAIN STRUCTURES - INSIDE FREEDOM

A regular grid of pillars holding up the floors of the different levels gives total freedom and the choice of internal division.
The stairwell, also made of reinforced concrete,
has the function of statically stiffening the whole structure, counteracting the horizontal forces (wind, earthquake).

188

189

为了保证室内温度的稳定性而使用了大量的沉重的玻璃幕墙，大楼水平和垂直方向都是用加固水泥建造的。

Because of the importance of the mass in fully glazed buildings to ensure the stability of the internal temperature, the horizontal and vertical structures have been made with reinforced concrete.

190

191

地热装置

地热装置管钻入地下达130米，共耗费2.5千米管道，水交换热泵其内耗低（消耗1千瓦时的电能可以换取4千瓦时的热能）。

GEOTHERMAL POWER PLANT

A system of geothermal probes, drilled into the ground up to a depth of 130 m for a total of 2.5 km of pipes, feeds a water-water heat pump that provides a valuable input of energy at low cost (for 1 kWh electricity consumed, 4 kWh of thermal energy is produced).

192

193

通风和空气调节

一套能供暖的通风系统能让室内空气保持一定的含氧量，用一种自然的方式消散沉积水分子，保证室内空气最优舒适度，减少因供热产生的能量消耗。

外部的空气通过建筑内的低压泵输入。由于地面收集器的作用，夏天空气输入前会预冷，冬天空气输入前会预热。排出的空气与输入的空气会进行充分的热交换，从而最大限度减少热量损失。

VENTILATION AND AIR CONDITIONING

A ventilation system with heat recovery maintains the right rate of oxygen in the air inside the building and eliminates the residual moisture in a natural way, ensuring optimal indoor comfort in every season and a substantial reduction in energy requirements for heating.
The fresh air taken from the outside is continuously pumped into the building at a low pressure.
Thanks to a ground collector, the air, before being introduced into the building is pre-cooled in the summer and pre-warmed in winter.
The outgoing air is made to intersect with the flow of incoming air in a heat exchanger with high efficiency: the resulting transmission of heat between the two flows minimizes any loss of energy.

194

195

太阳能收集系统——来自太阳的能量

朝南的墙面完美铺贴了100平方米的太阳能收集板（只有专业人员才看得出这是太阳能收集板），为大楼供应免费的热水。

太阳能收集板为取暖用水和卫生用水的加热起到了很重要的作用。

SOLAR COLLECTORS - ENERGY FROM THE SUN

In the south facade, 100 square meters of solar panels have been perfectly integrated (thus being recognizable in their function to a trained eye only), producing hot water for free.
The solar collectors contribute significantly to the production of hot water, both for heating and sanitary use.

196

197

光电伏系统——来自太阳的能量

光电伏系统集中在房顶，产生的电能供应大楼的整个制冷制暖技术操作系统的运行，以及普通照明（年需求量为18kWh/m²）。

PHOTOVOLTAIC SYSTEM - ENERGY FROM THE SUN

The photovoltaic system integrated into the roof produces the electrical energy required for the operation of all thc technological systems for heating, cooling and for lightings of the common spaces (total annual demand of 18 kWh/m²).

198

199

家庭自动化

产生能量的所有系统都有“串联式”功能：根据能量的需求，耗能低、成本低、效率高的系统最先运行。
家庭自动化系统可以通过触屏的 iPhone手机，iPad等全面控制建筑内运行的各种系统、装置，从而控制整幢大楼。
灯光、调光、音乐、录像、供暖、照相机这些装置都能实现远距离实时测控。
DeltaZERO 还设计了一套简单、雅致、方便理解的图形界面。

HOME AUTOMATION

All systems for the production of energy are based
on a “cascade” function: they are run by home automation,
and the systems with a more favorable cost / energy ratio
are made to run first, depending on the energy needs.
The home automation system allows users full control
of the various devices running the building
from touch-screens, iPhone and iPad.
Light, dimming, music, video, heating, cameras:
each device can be checked quickly and easily,
even at a distance and in real time.
The development of the graphical interface, very simple,
elegant and easy to understand,
was conceived by DeltaZERO.

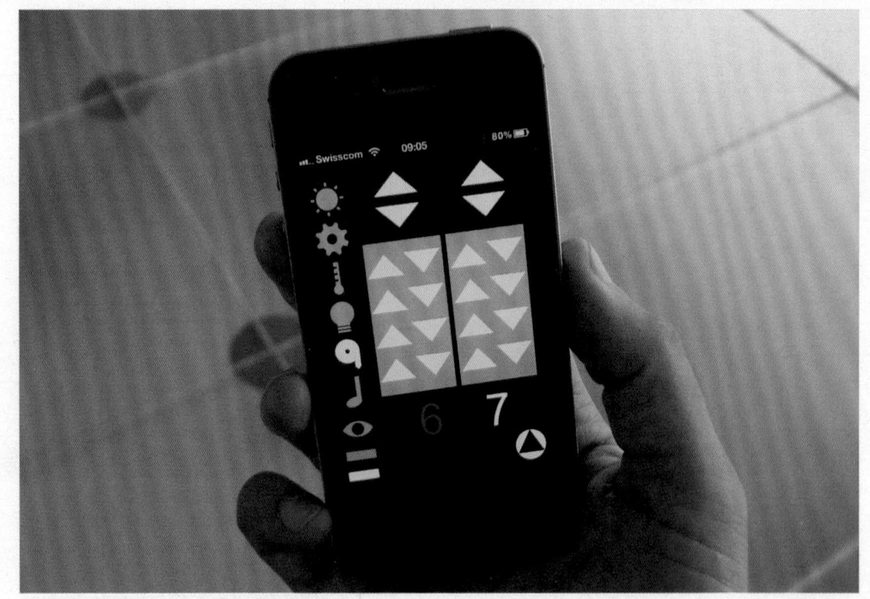

200

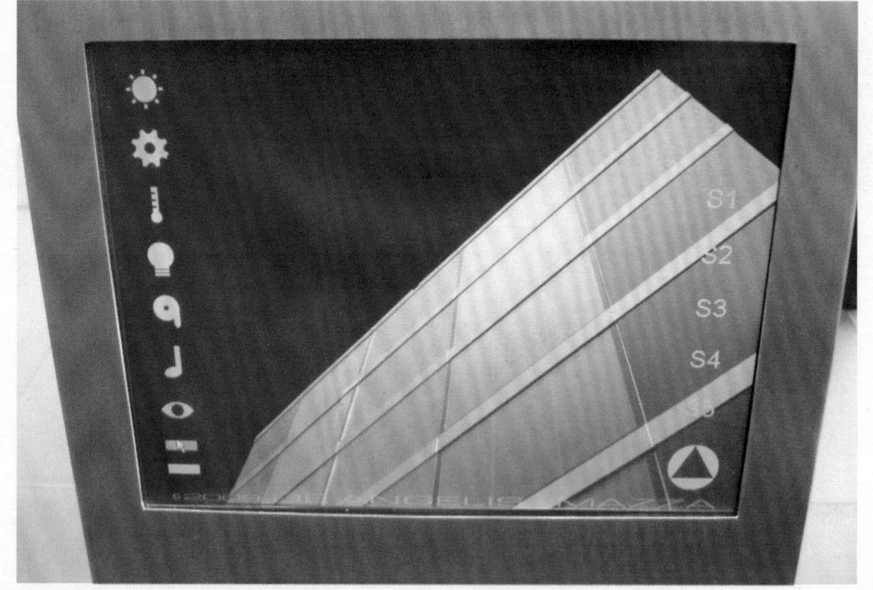

201

建筑外墙——设计

玻璃幕墙由3.20m × 2.60m ，无框隔热玻璃构成。
玻璃幕墙的分布特点会根据不同季节，天气的热量分布情况而改变。
每块隔热玻璃板厚63mm，并有充满氪气的双层孔，尽管光线可以穿透但是具备很好的隔热效果。

BUILDING SHELL - DESIGN

The glass facades are made of 3.20 m x 2.60 m glass plates with high insulating capacity and frameless.
The characteristics of each glass type was determined by simulations on the thermal behavior of the building in different seasons and weather conditions.
Each insulating glass plate has a total thickness of 63 mm, a double cavity filled with krypton gas and exceptional characteristics regarding insulation to heat despite the high light transmission.

202

203

建筑和艺术

DeltaZERO大楼简单的线条堪称建筑蕴含艺术的典范。
在大楼里Maria Mazza的浮雕和壁画环绕，人们或办公议事，或翩翩起舞，整幢建筑就象是充满生气的辉煌宫殿，时刻向窗外拥挤逼仄空间里的人们展示着玻璃幕墙后大楼里的惬意生活。

ARCHITECTURE AND ART

The simple linearity of the DeltaZERO building is a perfect setting for the inclusion of art in architecture. The characters that discuss or dance in the reliefs and murals of Maria Mazza are intended to give life day and night to the palace, reminding those who come in front of the externally compact and austere volume of the life that takes place inside the building behind the large square glass plates.

204

205

206

DeltaZERO大楼里的生活—— 一种独特的体验

大楼里空间完全开放，明亮、灵活、既可以居住也可以办公，还可以作为举办一些特殊活动的场所。
顶楼的液压平台可以在六楼和七楼之间运送人群，大型货物（搬迁、艺术展览、音乐会等等）。
带供热的低压通风系统会不断地为室内输送洁净新鲜的空气。空气在进入大楼前就经过了过滤电离处理。
系统会将空气中的含氧量与外部空气中的含氧量调节一致。
从空气质量这方面来说，住在DeltaZERO大楼就像是住在山林中，不管是在瑞士又干又冷的冬天还是湿热的地中海夏天，人们在大楼里生活都觉得舒适清新。

LIVING IN DeltaZERO - A UNIQUE EXPERIENCE

The space is completely open, bright and flexible, and is used as a residence, for work and often even for special events.
A hydraulic platform in the attic allows the transport of people and bulky items for special occasions (removals, art exhibitions, concerts, etc.) between the sixth and the seventh floors.
The ventilation system at low pressure and with heat recovery delivers clean and fresh air constantly in the rooms.
Before entering the building, the air is filtered and ionized.
The system provides a level of oxygen in the air equal to that of outside.
From the point of view of air quality, living in DeltaZERO is like being in the mountains; the living comfort is very good, both in the dry and cold Swiss winter and during the hot and humid Mediterranean summer.

207

208

合作建筑

DeltaZERO大楼是各方合作努力的成果。
不论从各个角度来看，DeltaZERO大楼都充分展示了这一特点。大楼的内部和外部都彰显它灵活、线性、富有设计感的风采。

CORPORATE ARCHITECTURE

DeltaZERO is a clear example of corporate architecture.
The building fully corresponds in fact even from a formal point of view to DeltaZERO's Corporate Identity.
Flexibility, linearity and design are radiated from the building both inside and outside.

209

210

211

碉楼生铁壶

碉楼生铁壶的设计灵感源于五邑地区的世界文化遗产开平碉楼，将中国建筑美学法则贯穿于生铁壶的各个细节，塑造出中国现代设计全新的视觉识别特征。

创新的泄气式解决了传统生铁壶热水喷溅问题，垂直收纳结构不仅节省放置空间，也赋予生铁壶这一传统产品更为时尚化的国际元素。这是一款原创的中国生铁壶，打破了日本生铁壶对中国市场的垄断格局。

碉楼生铁壶是近十年来唯一获得美国工业设计卓越奖（IDEA）碉楼生铁壶设计灵感源于五邑地区世界文化遗产开平碉楼，将中国建筑美学法则贯穿于生铁壶的各个细节，塑造出中国现代设计全新的视觉识别特征。创新的泄气方式解决了传统生铁壶热水喷溅问题，垂直收纳结构不仅节省放置空间，也赋予生铁壶这一传统产品更为时尚化的国际元素。这是一款原创的中国生铁壶，打破了日本生铁壶对中国市场垄断格局。

碉楼生铁壶是近十年来唯一获得美国工业设计卓越奖（IDEA）的生铁壶。IDEA是全球获奖难度最大的工业设计奖项之一，碉楼生铁壶的获奖彰显了中国原创设计及中国文化的巨大潜力。

"DIAOLOU" CAST IRON POT SETS

The inspiration of the pot design stems from the building Kaiping Dia lou (multi-storeyed defensive village houses in Kaiping, Guangdong province) which is one of the world's tangible cultural heritage.
Chinese architectural aesthetic principles can be easily discerned in each detail of the design, reflecting a completely new visually recognizable Chinese modern design feature.
The creative vapor venting system ensures that the boiled water will not spill over and cause any scalding.
The pot body featured by straight vertical lines claims less room to stack and displays an international fashion sense.
The original Chinese design pot is to pose a rival to the Japanese cast iron pot which has dominated the market for a long time.
Diaolou pot design winning the IDEA, an authoritative and high honor, undoubtedly demonstrates the great Chinese original design potential and the deepness of Chinese culture.

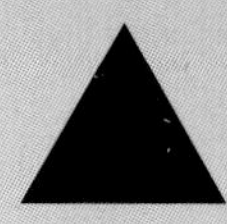

开平碉楼简介

鸦片战争之后，五邑地区的人开始离开故土，前往北美等地谋生，美国太平洋铁路修建期间，征用的大量华工，其绝大多数为五邑地区人士，五邑地区人的勤劳和智慧深深打动了美国太平洋铁路的修建方。部分华侨积累了第一批财富后回乡置房置业，这掀起了开平碉楼修建的历史。

抗日战争爆发后，大多数的碉楼主人回到北美，大多数碉楼被废弃。2007年碉楼成功申报为世界物质文化遗产，碉楼再次进入人们视野，每年有大量游客参观碉楼，碉楼成为了人们了解五邑文化的一个窗口。

碉楼生铁壶的造型源点源于开平碉楼，探索了碉楼保护的活态模式。

A BRIEF INTRODUCTION OF KAIPING DIAOLOU

In the 19th century the people who lived in Wuyi county had to leave their hometown because of the Opium Wars , and sought their living overseas. Most of them went to North America constructing the Pacific Railroad.
With hard work and wisdom, Wuyi people made their little fortune and some of them chose to come back to their hometown and built defensive houses to settle down. This is the start of Diaolou constructing history.
When the War of Resistance Against Japanese Aggression broke out, many Diaolou owners returned to North America, leaving the houses behind them deserted. In 2007 Diaolou successfully entered the world heritage list, and was opened to the public. Many tourists are attracted to visit Diaolou building and know more about Wuyi culture.

草图构思

1. 圆顶式壶盖及平台式壶身体现出中国传统认知中“天圆地方”的概念。

2. 壶身的生产制造采用中国传统的生铁铸造技术，表面处理采用中国的生漆技术。

3. 壶身顶部的围栏设计高低不平、错落延续，体现出中国山水的意境；壶身顶部的平台灵感来自中国建筑文化中“庭院”的概念。

SKECTH

1. Dome-formed pot lid and cubic body representing the ancient Chinese understanding of the universe "Hemispherical dome".

2. We employ the traditional Chinese casting technology to make the pot body, and raw lacquer techniques to process the appearance.

3. The uneven wall on top of the pot body resembles to a continuous stretch of mountain peaks, a conception featured by Chinese ink pictures.

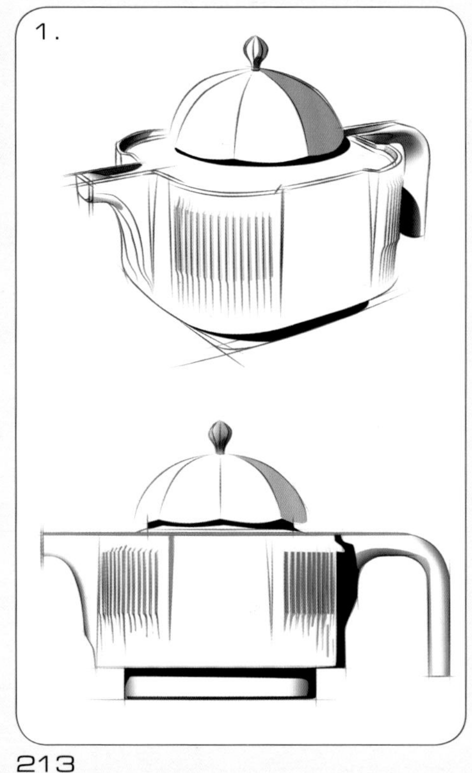

213

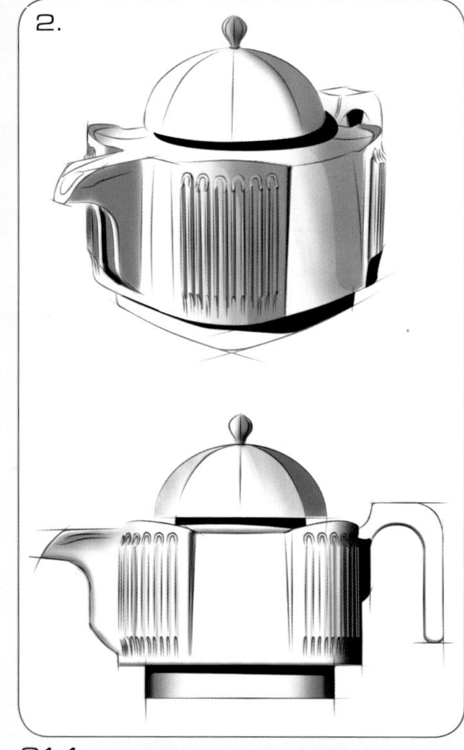

214

215

功能

文化创意产品与功能相结合才是一个好设计。在设计过程中，设计者发现了普通生铁壶热水易喷溅的问题，于是设计了一个非常创新的壶盖解决方案，这也是碉楼壶的最大设计亮点。

FUNCTION

To combine the cultural product with the function is a way to achieve a good design.
We find that it is easy for the traditional cast iron pot to scald the users because the boiling water may escape.
We designed a creative lid to solve the problem, which is also an attractive point in the Diaolou Pot.

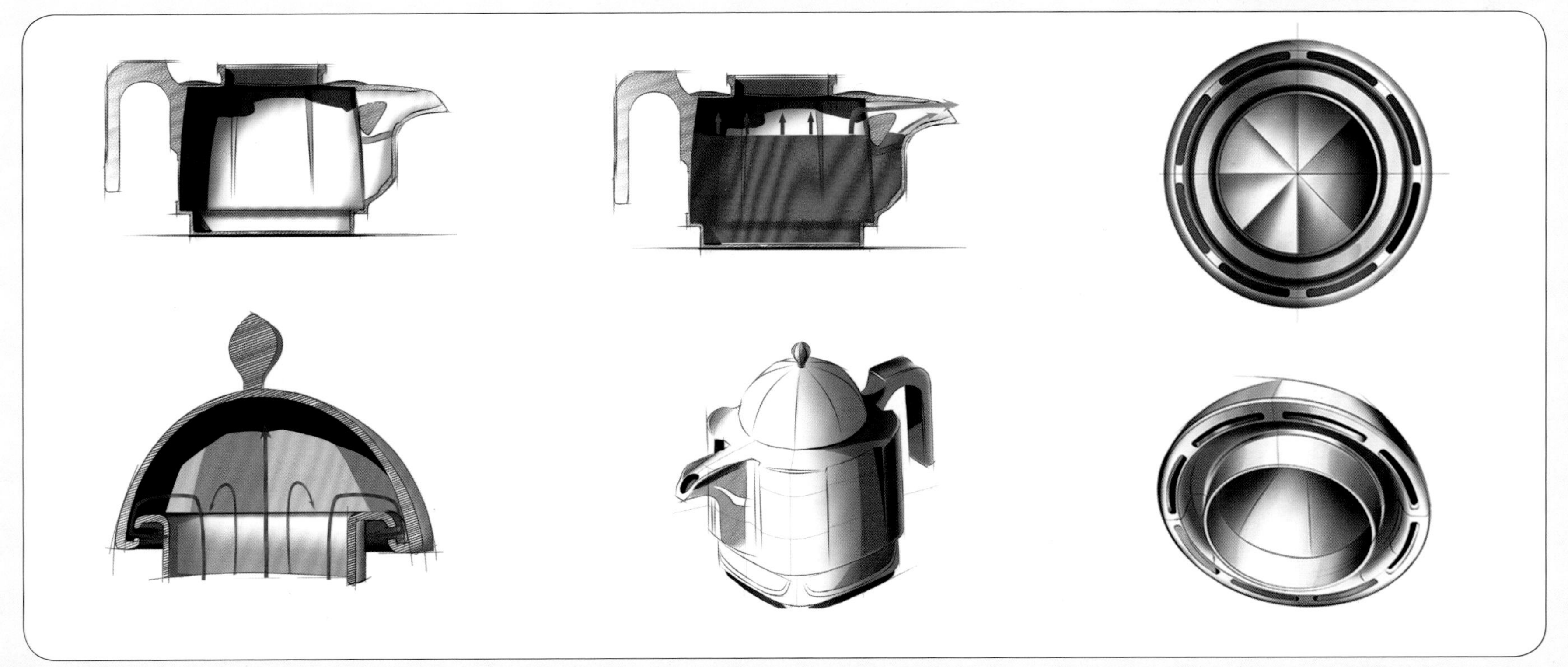

217

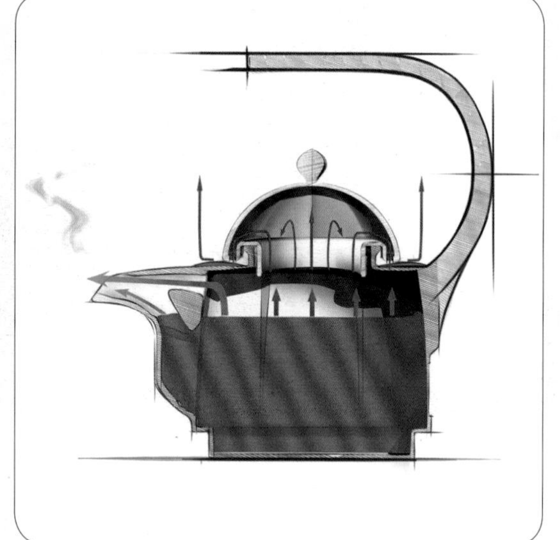

218

1. 高蒸汽压

对于传统的生铁壶，壶壁比较厚，不易导热。此外，传统生铁壶缺乏有效的蒸汽排放系统，当水煮沸时，由于蒸气压高，热水容易喷出口。

2. 低蒸汽压

该设计有8个创新的排汽口，提供了一个有效的排汽渠道，蒸汽容易溢出。另外，壶的壶口也具备排汽功能。新设计有效地防止了喷水。

1. HIGH VAPOR PRESSURE

Regarding the traditional cast iron pot, the wall of the pot is quite thick, which does not conduct heat outside easily. Plus, the pot lacks the effective vapor venting system.
So, when the water gets boiled, the hot water easily spurts out from the mouth because of the high vapor pressure.

2. LOW VAPOR PRESSURE

The design has 8 creative invisible slots on the brim of the lid that provides an effective channel for the vapor venting. Plus, the pot has a special vapor hole above the water hole. The mouth of the pot is the water channel as well as the gas channel. The whole design stops the water spurting out well.

节省空间

垂直收纳结构是设计该设计的另一个创新点，碉楼的垂直化布局有别于中国传统建筑的水平化布局。将壶、水洗、壶承及8个杯子垂直收纳，节省了近80% 的空间。

SPACE SAVING

To collect all parts vertically is another creative point in the design.
The layout of the Diaolou is arranged in a vertical way, which is quite different from the Chinese traditional tea set that is arranged in a horizontal way.
It saves 80% space to collect 1 pot, 1 pot holder, 1 cup holder and 8 cups vertically.

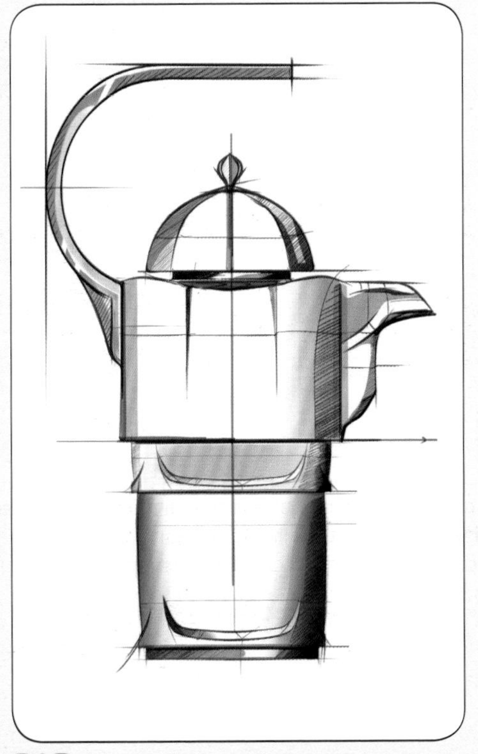
219

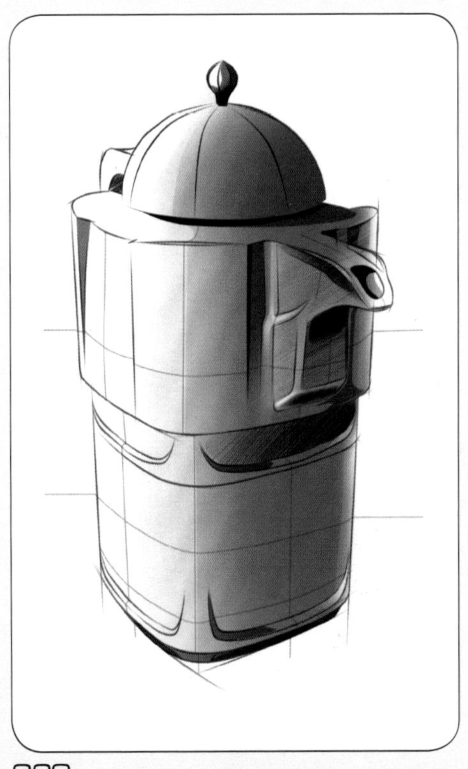
220

221

“圆满”杯子

底部为方形，顶部为圆形，在加水至70% 时，水面变为圆形。7成就是圆满，这是中国传统哲学的精髓。

"PERFECT" CUP

The cup has a square bottom and a round shaped top brim. To add tea to the 70% capacity of the cup, the shape of the water surface turns to round.
According to Chinese philosophy, round shape means perfection.
When someone accomplishes 70% of something, it is perfect. Do not go too far.

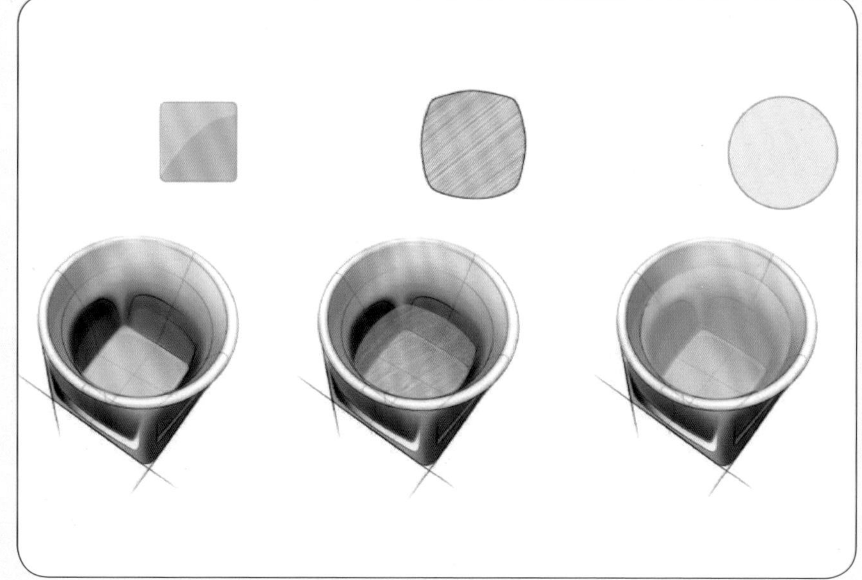

222

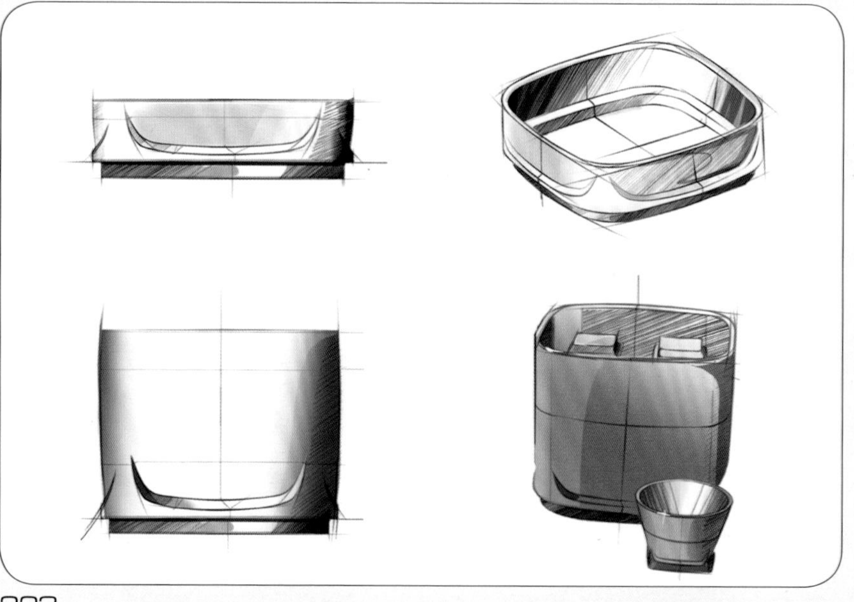

223

射蜡

因为碉楼壶的造型有别于传统生铁壶，其造型的高品质要求必须采用精密铸造的工艺，因此，我们加工多套模具以制作相应的蜡胚。在射蜡的车间需要控制良好的温度，以保证蜡胚不变形。

修蜡

将多个的蜡胚粘接在一处，制成一个完整的生铁壶蜡胚。

MOULDING THE WAX BASE

Because the shape of the Diaolou pot is different from the traditional pot, it needs to adopt the precise lost-wax casting technology to achieve the high quality appearance of the pot. We made several mounding tools to produc the wax base. In the wax base moulding shop, the room temperature needs to be controlled well, otherwise the wax base will be deformed easily.

ASSEMBLE THE WAX PARTS

Glue the all wax parts to make a complete pot base.

224

225

腊模组树

将完整的蜡胚壶粘接上铁水的流道及浇铸口。由于把柄与壶身为一体成型，因此浇铸口的位置非常高，这是对铸造工艺的一次挑战。

制壳

首先在蜡胚上裹上一层细沙，以保证产品的表面效果，然后再裹上一层粗沙浆以形成一个结实的外壳。

226

ASSEMBLING THE WAX BASE TREE

Adhering the liquid iron runner and the casting gate to the complete wax base.
Because the handle of the pot should be casted with the pot body as a whole one, the position of the casting gate is quite high, which is a real challenge for the lost-wax casting technology.

MAKING THE SHELL

Firstly it is necessary to wrap the wax base by the tinny sand to assure a perfect surface finishing.
Than we wrap the base with a rough mortar to make a tough shell.

227

风干

将模壳外表的粗沙浆风干，以使模壳完全定型。

高温高压脱蜡

将模壳放入烤炉中，浆蜡胚液化，形成空心模壳。

DRY BY WIND

Dry the shell by wind, the shell will be finalized completely, when the mortar is dried.

LOSING OF THE WAX

Losing wax in high temperature and high air pressure. Put the shell in the oven so the wax base will be liquified. Then a holly shell is formed.

228

229

模壳焙烧

将模壳放入高温炉中，烧制完全硬化的模壳。

BURNING THE SHELL

Put the shell in the high temperature oven and burn it to make a hard shell.

浇铸

采用攀枝花的钒钛生铁浇铸生铁壶，另外需要控制好浇铸的温度，以保证产品品质。

CASTING PROCESS

We cast the pot by using the vanadium titanium iron produced by PanZhiHua.
It needs to control the casting temperature well to assure the good quality product.

230

231

美国工业设计奖评语

生铁壶在亚洲有特别的文化意义，但在烧水时有烫伤人的危险，另其存放空间较大。碉楼生铁壶将吸引新一代的生铁壶使用者，因为该设计将传统的器具变得更安全，更节省空间。

碉楼生铁壶设计了创新的出汽方式，有效防止热水喷溅而出，另外，其巧妙的存放方式节省了收纳空间及时间。

IDEA REVIEWS

Cast iron pots have long held cultural significance in Asia, but posed a risk of scalding to the user from boiling water that may escape while the pots are being used,
and frustration with the amount of space they take up in the kitchen.
Now, Diaolou is attracting a new generation of cast iron pot users by making the traditional cookware safer and more space saving.
Diaolou cast iron pots have a creative vapor venting system to prevent hot water from spilling over.
They also stack easily to save space and time.

232

233

Stefano de Angelis-Effrem

Stefano de Angelis-Effrem (SdA)毕业于苏黎世联邦高等工业大学（ETH），毕业后在瑞士开办了自己的建筑设计公司。工作的同时他每年在米兰理工大学做系列建筑技术讲座。20世纪90年代他对城市规划和行政职能颇感兴趣，并在瑞士和意大利市政部门工作。

SdA在瑞士军队训练的经历让他拥有很强的组织能力，2001年掌管BAT G9，成为一名瑞士-意大利工程师营长。

他为私企和公有公司提供服务，也曾是国际银行的专家代表。他在欧洲和中国做关于未来时代的建筑以及城市及地域的可持续发展的讲座。他同时受聘为河南理工大学兼职教授。

作为DeltaZERO公司的一员，SdA在房地产建设，民用住宅规划建设，工、商业建筑，以及城市规划方面都非常活跃。

Maria Mazza

Maria Mazza (MM)毕业于米兰理工大学建筑专业。她在意大利北部多个建筑公司任过职。在银行业多年的工作经验让她能更好地了解客户需求。

她与de Angelis合作在米兰开办了de Angelis-Mazza建筑公司。她受聘为河南理工大学兼职教授。MM的主要工作是商用、民用建筑的规划及内部装修设计。她的专长是修复及具像艺术。她的艺术作品在很多展会和建筑里展出。

2003年，Stefano de Angelis和Maria Mazza建立了DeltaZERO公司，这是一个多学科共同规划组织，致力于未来时代的城市建筑设计，以人为本，恪守职业道德，朝同一个目标努力。该组织的宗旨是：

· 零能源消耗
· 空间布局及使用灵便
· 采用最先进的技术
· 锐意的现代设计感

温为才

1979年出生于江西赣州，毕业于北京理工大学，获工业设计方向博士学位。五邑大学副教授，“艺点禅心”品牌创始人。

2009年，与西班牙设计Jordi Mila合著《欧洲设计大师之创意草图》一书，获2009年度中国出版工作者协会输出版权优秀图书奖。2012年，出版著作《欧亚优秀工业设计案例透析》，作品“共生”被意大利现代艺术中心Arte Sella收藏。2013年，受邀参加国际米兰设计周国际绿色设计展举办个人设计展览，作品“葵艺灯具”获得了米兰设计周Goodesign证书。2015年，受到米兰世博会官方邀请，著作《产品造型设计的源点与突破》作为意大利与中国设计文化交流成果，在米兰世博会文化交流中心展出并做主题报告。2016年，碉楼生铁壶获得美国IDEA奖。2017年，作品“轮回”生铁壶、“葵艺+红木”灯具获得美国IDEA奖。

译者：陈华

2007年毕业于厦门大学外文学院英语语言文学专业，获硕士学位。主要从事英语翻译工作,翻译了欧洲著名汽车设计专家Enrico Leonardo Fagone的著作《汽车设计》。

Stefano de Angelis-Effrem

SdA graduated with a degree in architecture from ETH Zurich and in the same year he opened his architectural office in Switzerland.
In parallel with his profession, he taught annual cycles of lessons in architectural technology at the Polytechnic of Milan.
In the 90's, an interest in urban planning and for the functioning of institutions led him to occupy political offices at the municipal level in Switzerland and in Italy.
As a pioneer officer, SdA acquired organizational experience at the training courses of the Swiss Army and the command of different companies; in 2001 , he took command of the bat G 9, the battalion of Swiss-Italian Engineers.
His activity includes valuations for private and public clients; he was appointed as an expert by one of the major international banks.
He lectures on next generation buildings and on the sustainable development of the territory and cities in Europe and China.
He is adjunct professor at the Henan Polytechnic University.
Currently, as a member of DeltaZERO, SdA is active in real estate recovery, in the planning and construction of residential, commercial and industrial buildings and in urban planning.

Maria Mazza

MM graduated with a degree in architecture from the Politecnico di Milano.
She worked with different architectural companies in northern Italy.
MM acquired knowledge on the needs of commercial clients by working several years close to the banking milieu.
With the opening of de Angelis-Mazza Architects Milano, she started her partnership with de Angelis Associates in Lugano.
She is adjunct professor at the Henan Polytechnic University.
MM's main activity is the planning of commercial and residential buildings and interior design.
She specializes in restoration and loves the figurative arts.
Her artworks are displayed in and on different buildings and exhibitions.

In 2003 Stefano de Angelis and Maria Mazza established DeltaZERO, an interdisciplinary planning group committed to the development of cities and buildings for the iGeneration, based on man and ethics, sharing the same soul and objective:

- zero energy consumption
- flexibility of space and use
- integration of the latest technology
- sharp contemporary design.

Stone Wen

Stone Wen was born in 1979 in Ganzhou city, Jiangxi Province, China. In 2014 he obtained the Doctoral degree in industrial design from THE Beijing Institute of Technology. He is a professor of industrial design at Wuyi University. In 2016 he established his own product brand AEZM. In 2009 he co-worked with Jordi Mila and published the book Practical Design from the Top European Design Studio-EDDA. The book won the "2009 Outstanding Overseas Copyright Award" granted by Publishers Association of China.
In 2012 the book Case Analysis on Excellent Eurasian Industrial Design was also considered a big success.
In 2015 he was invited officially to present his book "The originality and the Break through of Product shape" at the EXPO Milano 2015. In 2013 he was invited to do his personal exhibition at the Milano Design Week 2013. In 2016 his design Diaolou cast iron pot set won the IDEA, which is one of the most prestigious prizes for design. In 2017 he got IDEA prize again by the Samsara cast iron pot and the Palm leaf & wood craft light design.

Chen Hua (translation e > cn)

Chen Hua graduated at the College of Foreign Languages and Cultures (CFLC) at Xiamen University (XMU) in 2007 and obtained the master's degree. Translator of "Car Design" written by Enrico Leonardo Fagone.

跋：

2012年应意大利现代艺术中心Arte Sella之邀，我携作品“共生”参加Arte Sella现代艺术展。

展览期间，见到各国大师的作品，其中Perry King（“红色情人节打字机”设计师）、Santiago Miranda（1992年西班牙塞维利亚世博会整体照明设计师）、瑞士建筑师Stefano De Angelis及其妻子Maria Mazza的作品都给了我很大的触动。

通过这些大师作品，我了解了西方设计师在阐述“自然、人、科技”的主题时的切入角度及思考过程。展览期间，我们一同度过了3天难忘的时光。

Stefano和Maria夫妇有瑞士人的理智、真诚，也有意大利人的幽默和热情。夫妇俩不遗余力地在全世界推广其“零能量”建筑的理念，难能可贵的是他们在欧洲以实际的建筑践行他们的理念，“知行合一”是这对夫妇给我的最深刻的印象。他们设计的瑞士保险大楼及“零能量”大楼都是世界上知名的建筑。中国代表团访问瑞士Lugano时也专门参观了“零能量”大楼。Stefano先生曾经执教于瑞士苏黎士理工学院，有过高校执教背景。作为一名大学教师，我觉得与他们特别投缘。

2014年及2016年，我们一同在中国大陆的一些高校举办过一些学术活动。Stefano与Maria向中国大陆的同行非常详细地介绍了“零能量”建筑及城市规划的基本理念、基本方法及一些设计实践。国内的同行都觉得将这样的理念推广到中国大陆可以为当下的中国建筑设计及城市规划提供一些借鉴和启发。于是我萌发了邀请Stefano和Maria夫妇为大陆设计界写一本书的想法。

在大陆，工业设计、建筑设计及城市规划设计是相对独立的设计领域，而欧洲设计大师的行业跨界是非常普通的。Stefano夫妇不仅是城市规划设计、建筑设计的行家，也是工业设计的行家，在现代艺术领域，夫妇俩在欧洲也有很深的造诣。2016年9月，我在罗马第一大学做讲座时，在瑞士的“零能量”大楼住过5天，该楼的技术的先进性，能量获得的创造性解决方案给我留下了深刻的印象。

写作的沟通过程是非常辛苦的，我们之间的往来邮件打印出来的厚度应该不亚于这本书的厚度。Stefano夫妇在写作过程中的严谨态度深深地感染了我。他们写作结构及表述内容的方式也给予了我很多启发。我也十分开心可以撰写这本书有关工业设计部分的内容。

此书得以在中国大陆首发，感谢电子工业出版社谭海平社长对本书的信任，感谢赵玉山编辑给予的巨大耐心，这一等就是3年的时间。感谢岑子健为本书中文版所做的辛苦排版工作。感谢陈华师姐的翻译工作，谢谢师姐对我一直的照顾。感谢南昌工程学院徐静婷老师为本书基础篇第4章所做的精彩翻译，你的翻译为本书增色不少。

感谢Stefano和Maria夫妇无私地将其数十年的设计经验及设计成果无私分享给中国的读者。今天正好是我女儿牛妞8岁的生日，今天可以为本书画一个圆满的句号尤为开心。

希望读者喜欢这本诚心之作。

温为才

2017年5月16日

Postscript :

In the end of 2012, I was invited by Arte Sella, the Modern art Center in Italy to do an exhibition. I brought my work "Gongsheng" to show the Western visitors. During the exhibition, I met Perry King, the designer of the Valentine Portable Typewriter, Santiago Miranda, who designed the lighting system for 1992 EXPR in Sevillay, Spain. And I met Stefano De Angelis and his wife Maria Mazza. The artwork from Maria is very inspiring for me. Through the artworks form the western designers, I understand the views and thinking processes when they explain the relationship of "nature, Human being, Technology". We together spent 3 very happy and days which is a very good memories in my life.

Stefano and Maria are rational and honest which is typical for the Swiss, while both of them are also humorous and hospitable likes most of Italian. The couple spread the theory of "Zero Energy" building, and they are practicing their own theory by the real building in Europe. They are the example of "Unity of Knowledge and Action" which impressed me very much. The Swiss insurance building and the "Zero Energy" building are world widely famous. When China delegation was visiting Lugano, they visiting the building. Stefano once taught in Swiss Federal Institute of Technology Zurich. As we have the same background, a university teacher, we are especially have a lot of things to share.

In 2014, 2016, we visited some universities and gave the lectures to the students. Stefano and Maria explained the basic theories, the methodologies and real practices to the designers and the students in mainland China. As designers, We think it helps the China architecture design practice, if the Zero- Energy theories could be introduced to China. So, I wanted to invited Stefano and Maria to read a book especially for China readers.

In Mainland China, the Industry design, architecture design and city planning are respectively independent fields. But in Europe, the role of designers are quite mixed in different field. Stefano and Maria are expertise in City planning and building design, they are also very good at industrial design. Fven in modern design, the couple are have their influences in Europe. In September 2016 I was invited by the First university of Rome to have a lecture to the western designers. I had been lived in Zero energy building for 5 day, in Switzerland, I was really astonished by the high technologies and the creative resolutions to save and achieve energy in the building.

It is really a very hard work to communicate during the writing process. If we print all the emails, I think the thickness of the emails papers is not less than the book. The professional and precise attitudes of the Stefano and Maria show the book are really moved me. The structure and the ways to express their theories in the book are also inspired me a lots. I am so happy can contribute and write some materials of the industrial design for the book.

The book could be published in Mainland China, I contributed it to the general manger of PHEI Tan Haiping, Tang trusts our group so much. Thanks to Zhao Yushan, You gave us the unbelievable patience in past 3 years. Thanks to Cen Zijian, thank you for your efforts for the layout work for the book in Chinese version. Many thanks to Chen Hua, as my older schoolmate, thank you for the translation work and for your kind help. Thanks to Xu Jingting, the teacher from Nanchang Institute of Technology your translation for the Chapter Prototype is very good.
Finally, a big thanks to Stefano and Maria. We are appreciating that you share your knowledge and experiences with the readers in China. Today is my daughter's birthday, named Alice, it is especially happy to make an end for the book.
Hope readers understand the efforts that we made for the book and like the book.

Stone 16th, May, 2017

Images and pictures - index.

01. Kristy, IMG_5451 - 15/01/2008 - www.flickr.com *
02. Kristy, IMG_5451-15/01/2008 - www.flickr.com *
03. European souther observatory, VTS snaps a very detailed view of the Triangulum galaxy - 06/08/2014 - www.flickr.com *
04. Ralph Buckley, Panel-6-Solar-System 14/02/2008 - www.flickr.com *
05. Kevin Dooley, Atom - 07/12/2013 - www.flickr.com *
06. JD Hancock, Riddle vs. Two-Face (66/365) 06/03/2010 - www.flickr.com *
07. Jaguar MEMA, C-X17 Sport crossover concept revealed, 09/09/2013 - www.flickr.com *
08. Jaguar MEMA, C-X17 Sport crossover concept revealed, 09/09/2013 - www.flickr.com *
09. The Car Spy, 2007 Rolls Royce Phantom, 11/09/2009 - www.flickr.com *
10. Jaguar MEMA, C-X17 Sport crossover concept revealed, 09/09/2013 - www.flickr.com *
11. Art Gallery Ergs-art, 27/08/2013 - www.flickr.com *
12. Scott J. Waldron, Knife under fork - 07/10/2006 - www.flickr.com *
13. Benny Lin, Chopsticks want you - 05/09/2006 - www.flickr.com *
14. Tim G. Photograpy, Telefono - 13/04/2013 - www.flickr.com *
15. Victor Svennson, The new Apple Iphone - 09/01/2007 - www.flickr.com *
16. Stefano de Angelis - Maria Mazza
17. (see 16. + 18.)
18. Ben, F12 Berlinetta - 05/07/2013 - www.flickr.com *
19. Marcus Kwan, Kobe Drawing - 03/04/2010 - www.flickr.com *
20. Intel free press, kids with educational computers 16/08/2013 - www.flickr.com *
21. Stefano de Angelis - Maria Mazza
22. Dick Thomas Johnson, Air Conditioner, 01/10/2014 - www.flickr.com *
23. William, Roman acqueduct x2 - 12/08/2012 - www.flickr.com *
24. Iosmininos, Sevilla, Puerte de ammalino - 09/09/2009 - www.flickr.com *
25. Martin Cooper, Red-bande sand wasp (ammophiila sabulosa) 01/08/2013 - www.flick.com *
26. (see 25. + 27.)
27. Fabbio, Chi vespa non mangia le mele - 01/05/2007 - www.flickr.com *
28. Russ2009, Matterhorn switzerland - 11/09/2012 - www.flickr.com *
29. Haldeman Brown, Toblerone - 01/03/2008 - www.flickr.com *
30. Jules, Toblerone ice cream cake-6 - 23/07/2011 - www.flickr.com *
31. Stefano de Angelis - Maria Mazza
32. (see 31. + 33.)
33. LC4 (1928) Le Corbusier, Pierre Jeanneret, Charlotte Perriand, Cassina 1965
34. Stefano de Angelis - Maria Mazza
35. (see 34. + 36.)
36. Abdullah AlBargan, P13_0023 - 27/11/2012 - www.flickr.com *
37. Stefano de Angelis - Maria Mazza
38. Stefano de Angelis - Maria Mazza
39. Riva aquarama - copyright: Archivio Riva (courtesy Ferretti Spa)
40. Stefano de Angelis - Maria Mazza
41. (see 40. + 42.)
42. Frank Derks, BMW headquarters - 22/09 2015 - www.flickr.com *
43. Gufram - Studio 65, Bocca - 1970 courtesy on Gufram srl - www.gufram.it
44. Edra, Tatlin - 1989 - courtesy of Edra spa - www.edra.com
45. b&b italia - Gaetano Pesce, Up5-6 - 1969 courtesy of B&B Italia – www.bebitalia.com
46. Unique Circle Yachts by Zaha Hadid Architects for Bloom+Voss Shpyards (visualisation Moka-Studio) - 2013
47. Zaha Hadid Galaxy Soho 014_Hufton & Crow
48. Zaha Hadid Architects, Nova shoes, 2013
49. Antjeverena, Outside-church of light - 09/07/2008 - www.flickr.com *
50. Chun-Hung Eric Cheng, Church of the light 17/10/2013 - www.flickr.com *
51. Stone Wen - Earphones
52. Achillli Family, Normandy American Chemetery, Omaha Beach, Colleville su mer , Normandia - 23/12/2014 - www.flickr.com *
53. Eric Huybrechts, German military cenerete WWII in Normandy18/08/2010 www.flickr.com *
54. Federico-Berti
55. David Orban, Chilometro Rosso - 15/10/2009 - www.flickr.com *
56. Lauren Manning, WoZoCo - 02/03/2008 - www.flickr.com *
57. Stefano de Angelis - Maria Mazza
58. Stefano de Angelis - Maria Mazza
59. Gnuckx, Hotel Ca' Sagredo - Grand Canal - Rialto - Venice Italy Venezia - Creative Commons by gnuckx - 30/05/2010 - www.flickr.com *
60. Wolfgang Staudt, The venetian - 27/07/2006 - www.flickr.com *
61. Bruno Kessler Marques, Porsche pavillion20/07/2012 - www.flickr.com *
62. Riley Kaminer, Old and New - 01/11/2009 - www.flickr.com *
63. Automobile Italia, Fiat 500 N del 1957 - 27/02/2015 - www.flickr.com *
64. Stefano, The new fiat 500 - 06/07/2007 - www.flickr.com *
65. Erik soderstorm, Containning the container #HDR #photog 13/02/2013 www.flickr.com *
66. Donna Jones, Freitag shop - bike display at entrance 18/10/2006 - www.flickr.com *

67. Verner Panton, Panton Chair - 1960s - copyright Vitra.
68. Tomas Gal, Chair - 01/01/2006 - www.flickr.com *
69. Alexey Vonokurov, Tron_lightcycle_wallpaper2 20/03/2011 - www.flickr.com *
70. The Community-Pop culture geek, E3 2010 disney booth lightcycle16/06/2010 www.flickr.com *
71. Stefano de Angelis - Maria Mazza
72. Georgi Nemtzov, Venus of Willendorf - 13/12/2009 - www.flickr.com *
73. Art Galler ErgsArt, Rubens Venus Frigid 1614 22/20/2015 - www.flickr.com *
74. Bea Serendipity, whati'm wearing today, 04/04/2012 - www.flickr.com *
75. Knoll Barcelona© Chair. Courtesy of Knoll.
76. Gavin Schaefer, Guggenheim MUSEUM - 01/03/2010 - www.flickr.com *
77. Da_morgado, dsc_2698 - 10/05/2015 - www.flickr.com *
78. Glenn Fleishman. Serra's wake, 26/05/2007 - www.flickr.com *
79. Bert Kaufmann, Cluod Gate Chicago (explore) 01/06/2008 - www.flickr.com *
80. Kinolamp, A dog and his boy, 14/07/2008 - www.flickr.com *
81. Airwolfhound, F117 - RIAT 2007 - 14/07/2007 - www.flickr.com *
82. Furtif evercut santoku. Courtesy of TB-groupe.
83. Benoit cars (ben), Black and love - 28/06/2012 - www.flickr.com *
84. Diana Olivares, Mezquita, 20/08/2013 - www.flickr.com *
85. Shella Thomson, Institut du monde arabe, 5/08/2006 - www.flickr.com *
86. Yann Caradec, Institute du Monde Arabe - 05/06/2014 - www.flickr.com *
87. Romain Winkel, Institute du Monde Arabe 4/10/2008 - www.flickr.com *
88. Julien Harneis, A school in waiting - 18/06/2007 - www.flickr.com *
89. Design Initiative, Jill Shaddock 'Splipcast' - 2/06/2011 - www.flickr.com *
90. Didrilks, Littala Teema - 24/08/2012 - www.flickr.com *
91. Brett Jordan, Carbon fibre 4 - 1/07/2010 -www.flickr.com *
92. Nick Cross, Reinforcement fabrics and matrials from Gurit 2/12/2008 - www flickr.com *
93. Steve Juvetson, Apollo 16 lunar module foil - 16/03/2012 - www.flickr.com *
94. David Bleasdale, dands3 - 24/02/2007 - www.flickr.com *
95. Didriks, Lekue silicone muffin molds - 30/09/2014 - www.flickr.com *
96. Martin Howard, Martini glasses - 7/08/2010 - www.flickr.com *
97. Stefano de Angelis - Maria Mazza
98. Steven Depolo, Cast iron anchor chains macro leland fishtoen macr 29/08/2010 www.flickr.com *
99. Ina D. Keating, The right fit - 23/08/2014 - www.flickr.com *
100. Jamie, Paper clip Wasteland-27/365 - 27/01/2010 - www.flickr.com *
101. Tony Hisgett, copper fittings - 18/03/2009 - www.flickr.com *
102. Max Klingesmith, Titanium - 26/03/2008 - www.flickr.com *
103. Bullion Vault, Whole gold bar on wood - 3/06/2009 - www.flickr.com *
104. Dennis Hill, fpx051913-09 - 14/05/2013 - www.flickr.com *
105. Frankieleon, if i could save time ina plastic bottle - 2/08/2011 - www.flickr.com *
106. Horia Varlan, Many colored straws thrown on top of each other 11/11/2008 - www.flickr.com *
107. Dave Cathpole, 25th ickworth wood & craft fair 2014 - 5/10/2014 - www.flick.com *
108. Gisela Francisco, clothespin - 4/12/2005 - www.flickr.com *
109. Alan Levine, Three spoons - 9/03/2010 - www.flickr.com *
110. Pete, Project 365 148:280513 First cut is the deepest 05/2013 - www.flickr.com *
111. Fagor Automation, Dnobat_Vertical_Lathe_cnc8065 - 05/06/2014 - www.flickr.com *
112. Beaartice Murch, Using the drill press - 15/08/2007 - www.flickr.com *
113. Savannah RIver SIte, Wader at SRS, 17/11/2009 - www.flickr.com *
114. Sam-Cat, Glue goo - 02/03/2009 - www.flickr.com *
115. LawPrier, Screw - 11/07/2009 - www.flickr.com *
116. SIgckgc, At the iron pour - 02/11/2014 - www.flickr.com *
117. Jeff Kubina - 02/10/2011 - www.flickr.com *
118. Derek Key, Blacksmith at work - colonial williamsburg - 30/10/2010 - www.flickr.com *
119. Dennis van Zuijiekom, vacuum formed '42' moulds - 13/07/2012 - www.flickr.com *
120. Roberto Cipriano, After bending - 23/07/2012 - www.flickr.com *
121. Dominic Alves, Chrome motorbike - 08/09/2001 - www.flickr.com *
122. Wikimedia commons, USMC-100818-M-4913M-001 .jpg - www.wikipedia.com *
123. George Smith - the architect's h and - 25/03/2009 - www.flickr.com *
124. Stefano de Angelis - Maria Mazza (Venezia)
125. Petrina Hu - Puerta tierra, Cadiz - 02/02/2008 - www.flickr.com *
126. Stefano de Angelis - Maria Mazza (Amman)
127. Nelson Minar - Houston Suburbs - 10/12/2010 - www.flickr.com *
128. Stefano de Angelis - Maria Mazza
129. Stefano de Angelis - Maria Mazza
130. Stefano de Angelis - Maria Mazza (Marrakesh)
131. Stefano de Angelis - Maria Mazza (Athina)
132. Stefano de Angelis - Maria Mazza (Paradiso)
133. Stefano de Angelis - Maria Mazza (Paradiso)
134. Joseysh Owaa - World class traffic jam - 23/04/2008 - www.flickr.com *
135. Tauno Tohk - Hong Kong 10/02/2010 - www.flickr.com *
136. Daviddje - London tube map in duch - 22/07/2012 - www.flickr.com *
137. Mikey - The “future” route master - 3/4/2013 - www.flickr.com *

138. Don Richards - Sandia Z accelelator - 12/09/2006 - www.flickr.com *
139. Black Rock Solar - Black rock solar Photovoltaic array at Pyramid lake high school - 19/01/2012 - www.flickr.com *
140. de Angelis - Mazza, Negrini, Torricelli - Lakefront Paradiso (competition)
141. La città vita - Pubblic space copenhagen - 02/02/2010 - www.flickr.com *
142. Daniel Inehner - Sky high - 22/08/2010 - www.flickr.com *
143. Kai Schreiber - Virtual head - 02/03/2007 - www.flickr.com *
144. Digitalpimp - Square off - 12/02/2011 - www.flickr.com *
145. Mohammed Shamma - Egyptian children of El-Fayoun - 24/12/2002 - www.flickr.com *
146. Dhani Borges - Favela 01 - 24/01/2010 - www.flickr.com *
147. Luke Ma - Cubic House - 05/05/2014 - www.flickr.com *
148. Stefano de Angelis - Maria Mazza (Oslo)
149. Seattle Municipal Archives - 1st Avenue S.Bridge Opening - 22/09/1956 www.flick.com *
150. Joi Ito - Executives tower, Dubay Business Bay - 20/02/2009 - www.flickr.com *
151. Maher Najm - High high... High enough to dream - 12/08/2015 - www.flickr.com *
152. Adnan Ghosheh - Burjalarab - 14/12/2011 - www.flickr.com *
153. Stefano de Angelis - Maria Mazza (Dubai)
154. Stefano de Angelis - Maria Mazza (Dubai)
155. Stefano de Angelis - Maria Mazza (Dubai)
156. Stefano de Angelis - Maria Mazza (Dubai)
157. Stefano de Angelis - Maria Mazza (Dubai)
158. Stefano de Angelis - Maria Mazza (Dubai)
159. Stefano de Angelis - Maria Mazza (Dubai)
160. Gavin - Palm 1 - 09/06/2006 - www.flickr.com *
161. Werner bayer - Atlantis the palm 2 - 28/10/2011 - www.flickr.com *
162. Stefano de Angelis - Maria Mazza (St. Gallen)
163. Stefano de Angelis - Maria Mazza (St. Gallen)
164. Stefano de Angelis - Maria Mazza (St. Gallen)
165. Stefano de Angelis - Maria Mazza (St. Gallen)
166. Stefano de Angelis - Maria Mazza (St. Gallen)
167. Stefano de Angelis - Maria Mazza (Kobenhavn)
168. Stefano de Angelis - Maria Mazza (Kobenhavn)
169. Stefano de Angelis - Maria Mazza (Kobenhavn)
170. Stefano de Angelis - Maria Mazza (Kobenhavn)
171. Stefano de Angelis - Maria Mazza (Kobenhavn)
172. Ermes Corti - Cantù - Piazza Garibaldi (courtesy of the author)
173. Stefano de Angelis - Maria Mazza
174. Stefano de Angelis - Maria Mazza
175. Stefano de Angelis - Maria Mazza
176. Stefano de Angelis - Maria Mazza
177. Stefano de Angelis - Maria Mazza
178. Luciano Carugo
179. Paolo Mazza
180. Stefano de Angelis - Maria Mazza
181. Stefano de Angelis - Maria Mazza
182. Stefano de Angelis - Maria Mazza
183. Stefano de Angelis - Maria Mazza
184. Stefano de Angelis - Maria Mazza
185. Stefano de Angelis - Maria Mazza
186. Stefano de Angelis - Maria Mazza
187. Stefano de Angelis - Maria Mazza
188. Stefano de Angelis - Maria Mazza
189. Stefano de Angelis - Maria Mazza
190. Luciano Carugo
191. Stefano de Angelis - Maria Mazza
192. Stefano de Angelis - Maria Mazza
193. Stefano de Angelis - Maria Mazza
194. Stefano de Angelis - Maria Mazza
195. Stefano de Angelis - Maria Mazza
196. Stefano de Angelis - Maria Mazza
197. Stefano de Angelis - Maria Mazza
198. Paolo Mazza
199. Stefano de Angelis - Maria Mazza
200. Stefano de Angelis - Maria Mazza
201. Luciano Carugo
202. Paolo Mazza
203. Luciano Carugo
204. Luciano Carugo
205. Luciano Carugo
206. Luciano Carugo
207. Paolo Mazza
208. Paolo Mazza

209. Paolo Mazza
210. Paolo Mazza
211. Stone Wen
212. Stone Wen
213. Stone Wen
214. Stone Wen
215. Stone Wen
216. Stone Wen
217. Stone Wen
218. Stone Wen
219. Stone Wen
220. Stone Wen
221. Stone Wen
222. Stone Wen
223. Stone Wen
224. Stone Wen
225. Stone Wen
226. Stone Wen
227. Stone Wen
228. Stone Wen
229. Stone Wen
230. Stone Wen
231. Stone Wen
232. Stone Wen
233. Stone Wen

* the original pictures were cropped and/or resized, and he colors were altered.

DESIGNING OBJECTS BUILDINGS CITIES
by Stefano de Angelis, Maria Mazza and Stone Wen

《设计的共性 —— 从产品、建筑到城市规划》由 Stefano de Angelis (SdA)、Maria Mazza (MM) 和温为才合著。书中图表资料由 Federico Berti和Alberto Cannizzaro 两位建筑师提供。

中文版式设计：瑞士“零能量”设计事务所，岑子健

本书内容出自：

- SdA在苏黎世联邦高等工业大学(ETH)获取的知识信息，特别是 Herbert Kramel 教授的系列讲座
- SdA 在米兰理工大学 (1991—1996) 关于建筑技术的系列讲座
- 著者们的实践经验

DESIGNING OBJECTS BUILDINGS CITIES
by Stefano de Angelis, Maria Mazza and Stone Wen

Collaboration (content, texts, graphics):
Federico Berti, Alberto Cannizzaro

Coauthor for the Chinese version: Stone Wen
Translator (English to Chinese): Chen Hua

This publication is based on:

- the knowledge gained by SdA at ETH Zürich and in particular focused on the lessons of prof. Herbert Kramel
- the lectures on tecnology in architecture made by SdA at Milan Polytechnic (1991-1996)
- the professional experience of the authors

 www.deltazero.net